Dissertations
in
American Economic History

This is a volume in the Arno Press collection

Dissertations in American Economic History

Advisory Editor
Stuart Bruchey

Research Associate
Eleanor Bruchey

See last pages of this volume for a complete list of titles.

STUDIES IN THE TECHNOLOGICAL DEVELOPMENT OF THE AMERICAN ECONOMY DURING THE FIRST HALF OF THE NINETEENTH CENTURY

Paul John Uselding

ARNO PRESS
A New York Times Company
New York – 1975

First publication in book form, Arno Press, 1975

DISSERTATIONS IN AMERICAN ECONOMIC HISTORY
ISBN for complete set: 0-405-07252-X
See last pages of this volume for titles.

Manufactured in the United States of America

Library of Congress Cataloging in Publication Data

Uselding, Paul J
Studies in the technological development of the American economy during the first half of the nineteenth century.

(Dissertations in American economic history)
Originally presented as the author's thesis, Northwestern University, 1970.
Includes bibliographical references.
1. United States--Economic conditions--To 1865. 2. Alien labor--United States. 3. Technology--United States--History. I. Title: Studies in the technological development ... II. Series.
HC105.U73 1975 330.9'73'05 75-2600
ISBN 0-405-07221-X

NORTHWESTERN UNIVERSITY

STUDIES IN THE TECHNOLOGICAL DEVELOPMENT OF THE AMERICAN ECONOMY DURING THE FIRST HALF OF THE NINETEENTH CENTURY

A DISSERTATION

SUBMITTED TO THE GRADUATE SCHOOL

IN PARTIAL FULFILLMENT OF THE REQUIREMENTS

for the degree

DOCTOR OF PHILOSOPHY

Field of Economics

by

PAUL JOHN USELDING

Evanston, Illinois

June, 1970

To

Mary Claire

ACKNOWLEDGMENTS

Professor H. F. Williamson, who first introduced me to the field of economic history, was most generous in sharing with me his gentle wit and wise counsel over many hours of discussion on ideas contained in the following chapters as well as on those ranging to the boundaries of the discipline.

My debt to Professor J.R.T. Hughes is beyond measure. He set the challenge and then showed the way. Through long talks and correspondence with him, my own views on economic history were shaped in countless ways. Admiration for the enormity and depth of his learning, and his sincere scholarly instincts permitted nothing less than total effort and commitment.

Professors F.P.R. Brechling and R. W. Clower provided invaluable help in tightening and strengthening the analytic arguments contained herein. Much of the material they critiqued does not appear in this thesis, but will emerge in future research.

Professor E. S. Ferguson provided encouragement throughout the course of the research and valuable help on substantive issues in the field of technological history.

A special debt of gratitude is due Dean R. H. Strotz whose foresight and understanding made all of this possible.

Paul J. Uselding

Urbana, Illinois

March 1, 1970

TABLE OF CONTENTS

Chapter | Page

LIST OF FIGURES

LIST OF TABLES

LIST OF TABLES (Continued)

LIST OF TABLES (Continued)

CHAPTER I

THE LABOR SCARCITY HYPOTHESIS AND TECHNOLOGICAL DEVELOPMENT: THE HABAKKUK-TEMIN-FOGEL DILEMMA

According to a substantial body of contemporary evidence the "dearness" of labor in America and the resulting high wage was responsible for the unique development of American technology. This is the well-known labor scarcity hypothesis first put forth by E. Rothbarth[1] and later elaborated by H. J. Habakkuk.[2] In an effort to dispel the notion that different capital-labor ratios in Britain and America implied different technologies, i.e., production functions, Peter Temin showed how such a divergence in capital-labor ratios could come about simply as a result of differing factor-price ratios between two countries using the same technology.[3]

Temin's startling conclusion that the American "real wage" was lower than Britain's, and that America was therefore a labor abundant economy resulted from a "specification error" in his model. Robert Fogel altered Temin's formulation in such a way as to deny the latter's conclusion that America was a capital scarce and labor abundant economy.[4]

Implications of the Temin and Fogel Models

This section presents the models specified by Temin and Fogel with a discussion of the conclusions drawn from those models as originally specified.

Temin's model used three factors of production and two sectors, manufacturing and agriculture. The manufacturing sector used only capital and labor, while agriculture employed only labor and land.

The two production functions were as follows:

$$(1) \qquad Q_M = g(L,K)$$

$$(2) \qquad Q_A = f(L,T)$$

where,

Q_A = the output of agriculture

Q_M = the output of manufacturing

L = labor

K = capital

T = land

These production functions are required to be homogeneous of degree one and to have positive first derivatives and negative "own" second derivatives. "In other words there are no economies or diseconomies of scale, factors of production are never redundant, and there are diminishing returns to any single factor."[5] Fogel has shown that corresponding to equation (1) above there is a function which is Cobb-Douglas and meets Temin's specification.

$$(3) \qquad Q_M = GL^{a_1}K^{a_2} \; ; \quad a_1 + a_2 = 1$$

Equation (3) can be used to derive a factor-price frontier, which is the factor-price dual of the production possibility curve. The frontier shows the maximum real wage that can be obtained with a given technology for any given interest rate. As is shown in the appendix to this chapter the factor price frontier corresponding to equation (3) is

$$(4) \qquad r = cP_M^{1/a_2} G^{1/a_2} w^{-a_1/a_2}$$

where,

$c = (a_1^{a_1/a_2} a_2)$

r = the money rental rate of capital

P_M = the price of Q_M

w = the money wage rate

G = an index of technical efficiency

Dividing (4) by P_M results in

$$(5) \quad i = cG^{1/a_2}\tilde{w}^{-a_1/a_2}$$

where,

i = nominal rate of interest, r/P_M

$\tilde{w}$ = w/P_M, or Temin's "real wage"

A graph of equation (5) would appear as in Figure 1.1.

Temin, using evidence from Sidney Homer, asserted that, "the interest rate in the United States was consistently higher than the rate in Britain during the first half of the nineteenth century."[6,7]

Given this evidence on interest rates, and the model he used, Temin was led to the conclusion via Figure 1.1 that if $i_a > i_b$ then $\tilde{w}_a < \tilde{w}_b$ and hence that America was labor abundant, i.e., the American "real wage" was less than Britain's. Fogel then showed how this conclusion resulted from a "specification error."

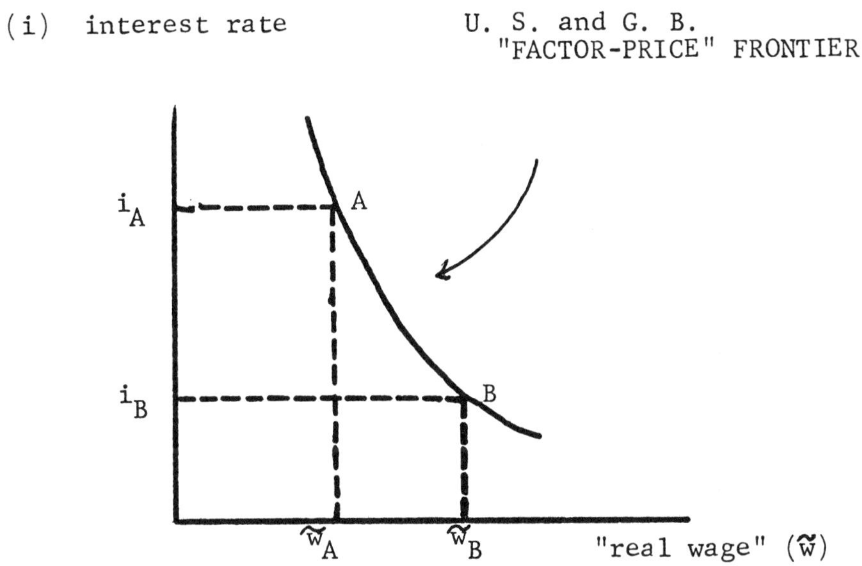

FIGURE 1.1

THE FACTOR PRICE FRONTIER

> Temin requires identical technologies in order to argue that a relatively higher American interest rate implies that its wage rate was relatively lower. Since Temin's evidence was inadequate to establish (this assumption) he has not eliminated the possibility that the United States was both labor- and capital-scarce. And this dual scarcity is the traditional position of economic historians.[8]

For the argument at hand Fogel is willing to grant the assumption of identical technologies between Britain and America. However he respecifies Temin's production function for the manufacturing sector by including the output of agriculture as a factor of production in the manufacturing sector.

$$(6) \qquad Q_M = GL^{a_1}K^{a_2}Q_A^{a_3} \; ; \quad a_1 + a_2 + a_3 = 1$$

This specification is quite plausible, as Fogel argues, since the products of farm, forest, and mine such as cotton, hemp, flax, lumber, and raw ores are inputs to the production process. This is merely to say that raw material is a factor of production. Corresponding to equation (6), Fogel then derives the following factor-price frontier.

$$(7) \qquad r = dP_M^{1/a_2} G^{1/a_2} w^{-a_1/a_2} P_A^{-a_3/a_2}$$

where,

$$d = \left(a_1^{a_1} a_2^{a_2} a_3^{a_3}\right)^{1/a_2}$$

P_A = price of agricultural commodities

Dividing both sides of equation (7) by P_M then yields.

$$(8) \qquad i = dG^{1/a_2} \tilde{w}^{-a_1/a_2} \tilde{P}_A^{-a_3/a_2}$$

where,

$\tilde{w}$ = w/P_M, Temin's "real wage"

$\tilde{P}_A$ = P_A/P_M

i = Temin's rate of interest, r/P_M

Dividing both sides of equation (5) by $\overline{P}$ yields:[9]

$$(9) \qquad i = cG^{1/a_2}\hat{w}^{-a_1/a_2} \left(\frac{P_M}{\overline{P}}\right)^{1/a_2}$$

where,

$i = r/\overline{P}$

$\hat{w} = w/\overline{P}$, Fogel's real wage

$\overline{P} = k_1P_M + k_2P_A$ = an index of commodity prices

k_1 = share of income spent on Q_M

k_2 = share of income spent on Q_A

Fogel derived equation (9) in order to get $\hat{w}$, the real wage $w/\overline{P}$, instead of $\tilde{w}$, the marginal physical product of labor, which is given on the right-hand side of equation (8). In doing so he has altered Temin's definition of the nominal rate of interest. The equilibrium rate of interest is usually described by the following relationship:

$$(10) \qquad r = p_k \cdot i \qquad \text{or} \qquad i = \frac{r}{p_k}$$

where,

r = money rental rate of capital

p_k = price of capital

i = equilibrium rate of interest

If $P_M = P_k$, then on Temin's specification of equation (5), i represents the equilibrium rate of interest, while $\tilde{w} = w/P_M$ is not the "real wage." Fogel's attempt to obtain the real wage $\hat{w} = w/\overline{P}$ in equation (9) has resulted in the definition of i as $r/\overline{P}$, which is not necessarily the equilibrium rate of interest. The term $r/\overline{P}$ is the real rental of capital. It would be the equilibrium rate of interest only if physical capital was immortal.

Temin's inference of labor abundance is then denied by appeal to Figure 1.2.

Figure 1.2 shows the U. S. factor price frontier everywhere above Great Britain's frontier as the result of using $(P_M/\overline{P})$ as a shift parameter. This result is obtained with an assumption of common technologies by analyzing the comparative prices of manufacturing and agricultural output in the two economies. If $(P_M/\overline{P})_{U.S.} > (P_M/\overline{P})_{G.B.}$, then by equation (9) the American frontier is shifted outward. That the relative price of U. S. manufactures should exceed the British price may be plausibly argued by anticipating the effects of local or regional monopolies in the U. S. fostered by high inter and intra national transport costs, tariffs, and factor immobilities. Abundant

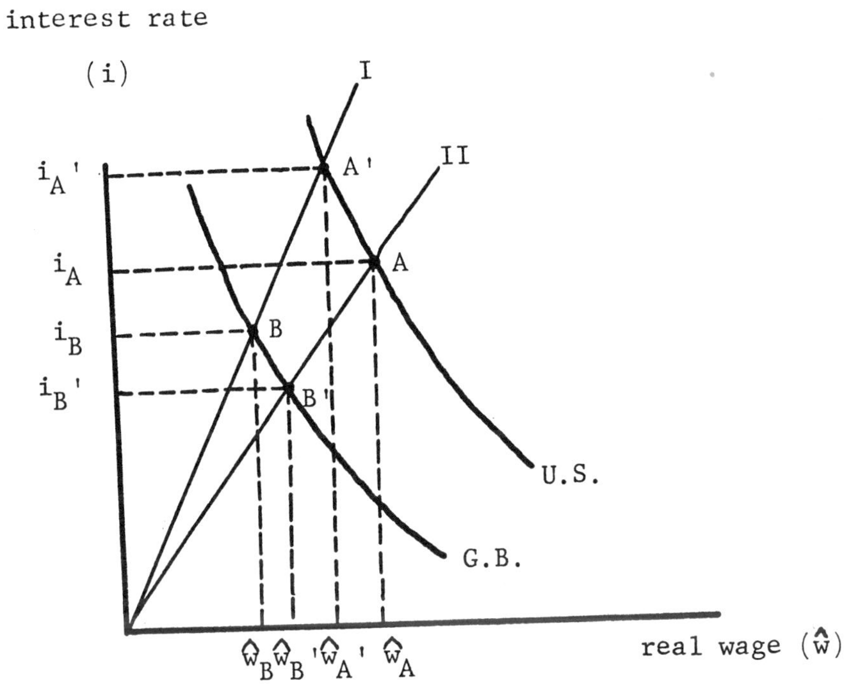

FIGURE 1.2

THE HYPOTHETICAL POSITIONS OF THE AMERICAN AND BRITISH FRONTIERS

and fertile land, by reducing the price of agricultural output in the U. S. relative to Britain, also raises the relative price of manufacturing output as may be seen from the following.

$$(11) \qquad \frac{P_M}{\overline{P}} = \frac{1}{k_1 + k_2 \left(\frac{P_A}{P_M}\right)}$$

If the American price of agricultural, relative to manufacturing, output is lower than the British price then the denominator of equation (11) will be smaller thus tending to raise the American ratio $(P_M/\bar{P})$ relative to Britain.[10]

With the frontiers as shown by Figure 1.2, then a higher American interest rate, i_a', also corresponds to a higher American wage rate, $\hat{w}_a$. Thus Fogel's model has temporarily reinstated the view he holds as "traditional," namely that the United States could have been both labor and capital "scarce." The slopes of lines II and I through points A and B, respectively of Figure 1.2 are

$$(12) \qquad \frac{i_a}{\hat{w}_a} = \frac{a_2 L_a}{a_1 K_a}$$

$$(13) \qquad \frac{i_b}{\hat{w}_b} = \frac{a_2}{a_1} \frac{L_b}{K_b}$$

As shown in Figure 1.2, when the slope of I exceeds that of II, and the U. S. is at A and Great Britain is at B, the U. S. is employing relatively more capital per worker than is Britain. Thus Fogel claimed to have reinstated the dual scarcity position by demonstrating that (1) both American real factor prices exceeded Britain's and (2) this could also imply a higher capital labor ratio in America as compared to Britain. This last point is simply the labor scarcity hypothesis of Rothbarth, Habakkuk _et al_.

The assumption of equal factor shares in the U. S. and G. B. also implies this result. Let

$$(14)\qquad (a_1/a_2)_{U.S.} = (a_1/a_2)_{G.B.} = \left(\frac{w.L}{r.K}\right)_{U.S.} = \left(\frac{w.L}{r.K}\right)_{G.B.}$$

Now if $(w/r)_{U.S.} > (w/r)_{G.B.}$ (where w = money wage rate, r = rental of capital), the direct implication of the use of a Cobb-Douglas function is $(L/K)_{U.S.} <$ $(L/K)_{G.B.}$ because the elasticity of substitution is unity.

If the slope of line I is allowed to represent the ratio of the "interest rate" to the real wage rate in America, and line II the same ratio for Britain, (i.e., the opposite of Fogel's suggestion), then the U. S. will be at A' with the real wage $\hat{w}_a'$ and "interest rate" i_a', and G.B. will be at B' with $\hat{w}_b'$ and i_b'. In this case $\hat{w}_a' > \hat{w}_b'$ and $i_a' > i_b'$ i.e. dual scarcity, but now the implication is that $(K/L)_{U.S.} < (K/L)_{G.B.}$. In other words the use of this factor price frontier model under the assumption of dual scarcity ($\hat{w}_a' > \hat{w}_b'$ and $i_a' > i_b'$) does not necessarily imply that American production was characterized by capital intensity relative to Britain. A higher American interest rate, under Fogel's specification could imply that the United States was either more or less capital intensive relative to Britain. In terms of the model, the question of relative capital intensity cannot be established by an ordinal estimate of the interest rates alone i.e., depending on whether the United States is at i_a' or i_a or Britain is at i_b' or i_b we would get entirely opposite conclusions.

In fact if the shift parameter $(P_M/\overline{P})$ is only infinitesimally larger for the United States relative to Britain then the American frontier will not be sufficiently above the British frontier in Figure 1.2 to reject Temin's position i.e. a higher American interest rate implies a lower American "real wage." Fogel assumed the relative magnitudes of the two shift parameters were disparate enough to put the frontier in the position shown by Figure 1.2, and this again implied knowledge of the cardinal values to be assigned to the variables. This same argument applies to Fogel's use of the shift parameter $\tilde{P}_A$ in equation 8, derived from his respecification of the manufacturing production function.

The Temin-Fogel model is efficient in that, in principle, the level real wages could be deduced if the interest rates were known; only if the specific levels were known with certainty however, could we accept the conclusion of relative, American capital-intensity. As has been shown, Fogel's re-instatement of the traditional position, can be reversed as interest rates are allowed to move to different hypothetical levels in the two countries, even though the United States remains in a position of dual scarcity. It is not sufficient to know that the American interest rate was greater than the British rate (an ordinal estimate). The question is how much greater (cardinal numbers)?

The Impact of an Historical Perspective

In view of some recent findings by Nathan Rosenberg,[11] the Temin-Fogel model may be respecified in a way that may be closer to historical reality. Rosenberg has brought to light new evidence on wage differentials between Britain and America. The principal finding for our purposes here is that in the United States machine makers and skilled labor received a smaller premium over the wages of common labor than in Britain. Taking the wages of common labor in each country as a base index of 100, and using Zachariah Allen's data,[12] Rosenberg shows that in the case of ordinary machine makers the range for the U. S. is 125 to 142 as compared to the English index of 149. In the category of "best machine makers" the U. S. index ranges from 150 to 175 while in England it is given as 262 or over two and one-half times the rate of compensation paid to common labor in the latter country. It was further found by grouping the observations on the thirteen occupational categories in Allen's table that for England,

$$\frac{\text{average wage index of skilled workers}}{\text{average wage index of unskilled workers}} = 208$$

The same ratio for the United States is 199. While the index used here is of an extremely rough sort, and the relative price differential of skilled and unskilled workers between the two countries is slight, Rosenberg's evidence is suggestive of an important consideration in approaching the question of labor scarcity. The labor scarcity hypothesis and the Temin-Fogel models deal with homogeneous labor, while in fact it may be the different types of labor employed in manufacturing and agriculture that are central to any understanding of how and why American technology and manufacturing practice diverged from the British experience. The relative price of skilled to common labor in America was less than in England according to Zachariah Allen. Since a particular type of skilled labor, i.e., the machine maker is used to produce capital goods, it is the price of this type of labor that is also pertinent to an analysis of capital-labor substitution. If Allen's figures are representative of the rate of compensation of skill differentials in America and England, then it is less costly to substitute machinery, embodying relatively cheaper skilled labor, for common labor in America than England, i.e., the premium for skilled over unskilled labor is less in America. If the labor services of machine makers were absolutely cheaper in America then the mechanism involved in the substitution of capital for unskilled labor is more readily apparent, i.e., the labor cost component of capital goods was lower in America than in England. Assuming the same factor proportions and technologies in the capital goods industry of each country, the cost of production of capital would be absolutely lower in the United States.[13]

Neglecting the effect of depreciation, we may show the effect of this suggestive evidence on the price of capital goods for the argument at hand.

Substituting equation (9) into equation (10) gives

$$(15) \qquad r = p_k \left[c \left(\frac{P_M}{\overline{P}} \right)^{1/a_2} G^{1/a_2} \hat{w}^{-a_1/a_2} \right]$$

Dividing (15) by $\overline{P}$, the index of commodity prices, yields:

$$(16) \qquad i = \frac{r}{\overline{P}} = \left(\frac{p_k}{\overline{P}} \right) \left[c \left(\frac{P_M}{\overline{P}} \right)^{1/a_2} G^{1/a_2} \hat{w}^{-a_1/a_2} \right]$$

or Fogel's "rate of interest" $(r/\overline{P})$. Now the question of whether or not the American factor price frontier lies above (or below) the English frontier turns also on the ratio of the index of the price of capital goods to the index of all commodity prices, $(P_k/\overline{P})$. If the ratio for the U. S. is less than the corresponding

British ratio, as has been argued above, the effect will be to shift the American factor price frontier closer to the origin. On the other hand Fogel has argued that

$$\left(\frac{P_M}{\overline{P}}\right)_{U.S.} > \left(\frac{P_M}{\overline{P}}\right)_{G.B.}$$

tending to shift the American factor-price-frontier outward. Thus whether Temin's conclusions, America was labor abundant by virtue of a lower real wage, or the traditional position, dual scarcity in America, as re-formulated by Fogel is tenable in part turns on the relative magnitudes of two offsetting shift parameters in the factor-price-frontier as represented by equation (16). The relative position of the frontiers for both countries depends on whether

$$(17) \qquad \left(\frac{P_k}{P_M}\right)_{U.S.} \gtrless \left(\frac{P_k}{P_M}\right)_{G.B.}$$

It would be necessary to get specific estimates of the ratios in equation (17) in order to establish the relative position of the two frontiers.

If equation (10) had included the effect of depreciation on the rental rate of capital

$$(18) \quad r = p_k(i+d')$$

where

d' = rate of depreciation

then the depreciation rates in the two countries would also have acted as a shift parameter in equation (16), further tending to loosen the simple connection between relative agricultural and manufacturing prices and the position of the frontier.

Ames and Rosenberg[14] have suggested that in the use of production functions for purposes of historical analysis it is appropriate to detail specifically the names of different types and kinds of inputs.

> Historians often have to consider real production processes. They must, for instance, replace the numbers 1, . . ., n by the names of inputs actually used in a given situation. It makes a difference, for example, whether one of these numbers stands for a hand chisel or a Blanchard gunstocking lathe.[15]

In the spirit of this suggestion we could re-specify equation (6) (originally Fogel's) as follows:

$$(19) \quad Q_M = G L_u^{a_1} K^{a_2} Q_A^{a_3} L_s^{a_4}$$

$$(a_1+a_2+a_3+a_4) = 1$$

where

L_u unskilled labor

L_s skilled labor

This way of writing the production function for the manufacturing sector simply admits that labor is non-homogeneous, and specifically two types, skilled and unskilled, are employed in manufacturing. Corresponding to equation (19) the factor-price-frontier would be given by

$$(20) \qquad i = e \left(\frac{P_M}{\overline{P}}\right)^{1/a_2} G^{1/a_2} \hat{w}_u^{-a_1/a_2} \left(\frac{P_A}{\overline{P}}\right)^{-a_3/a_2} \hat{w}_s^{-a_4/a_2}$$

where

$$e = \left(a_1^{a_1} a_2^{a_2} a_3^{a_3} a_4^{a_4}\right)^{1/a_2}$$

If we wish to graph the frontier of the "interest rate" against the real wage for unskilled labor we have three shift parameters in equation 20 assuming common production functions, i.e., $(P_M/\overline{P})$, $(P_A/\overline{P})$, and $\hat{w}_s$. The effect of all three of these would be to shift the American frontier everywhere above the British one given Fogel's argument and Allen's suggestion that $(w_s)_{U.S.} < (w_s)_{G.B.}$ If instead we use $(P_M/\overline{P})$, $(P_A/\overline{P})$, and $\hat{w}_u$ as shift parameters while graphing a relation between i and $\hat{w}_s$, then the relative position of the frontiers for the two countries is not known since

$$(\hat{w}_u)_{U.S.} > (\hat{w}_u)_{G.B.}$$

that is the wage of unskilled labor in America is greater than in England, as is suggested by Allen's data. On this last specification the relative magnitudes of $(P_M/\overline{P})$ and $(P_A/\overline{P})$ will tend to shift the American frontier everywhere above the English one; however if the assumed condition for unskilled wages holds this would tend to shift the American factor-price-frontier inward toward the origin, offsetting the effect (to some unknown extent) of the other two shift parameters. Once again, depending upon the magnitude of these relative effects neither Temin's or Fogel's conclusions can be upheld given the present state of empirical knowledge. However, under the above specifications it is extremely unlikely that the frontier for the two countries would coincide precisely so as to reinstate Temin's position. On the other hand it is possible that Fogel's specification leading to the conclusion that the American frontier was everywhere above the English frontier may also be incorrect.

If we assume that only unskilled labor is used in agriculture then the agricultural production function would be as follows:

$$(21) \qquad Q_A = BL_u^{b_1}K^{b_2}T^{b_3}, \quad b_1 + b_2 + b_3 = 1$$

where

Q_A = the output of agriculture

L_u = unskilled labor

T = land

Then the factor-price-frontier showing the rate of interest and the real wage would be

$$(22) \qquad i = g\left(\frac{P_A}{\overline{P}}\right)^{1/b_2} B^{1/b_2}\hat{w}_u^{b_1/b_2}\hat{1}^{-b_3/b_2}$$

where

$g = \left(b_1^{b_1} b_2^{b_2} b_3^{b_3}\right)^{1/b_2}$

i = rate of interest

B = index of technological efficiency in agriculture

$\hat{w}_u$ = real wage of unskilled labor

$\hat{1}$ = real rental of land

P_A = price of agricultural output

$\overline{P}$ = index of commodity prices

In graphing the relation between i and $\hat{w}_u$, $\hat{1}$ and $P_A/\overline{P}$ may be used as shift parameters. Since

$$(23) \qquad \frac{P_A}{\overline{P}} = \frac{1}{k_1\left(\frac{P_M}{P_A}\right) + k_2}$$

the larger quantity of land in America would cause (P_M/P_A) for America to be higher than in Britain. Thus

$$(24) \qquad \left(\frac{P_A}{\overline{P}}\right)_{U.S.} < \left(\frac{P_A}{\overline{P}}\right)_{G.B.}$$

tending to shift the American frontier closer to the origin. Land rental may be expressed as

$$(25) \qquad 1 = P_A \cdot MPP_T$$

where MPP_T = marginal physical product of land. Since we assume $(P_A)_{U.S.} < (P_A)_{G.B.}$, this by itself would make $(1)_{U.S.} < (1)_{G.B.}$. However, if $(MPP_T)_{U.S.} > (MPP_T)_{G.B.}$, a not unreasonable assumption due to the fertility of soil in the American Mid-West, then this component of the land rental equation would tend to offset the effect of the lower price of agricultural commodities for the United States in equation 25. Thus whether

$$(26) \qquad (\hat{1})_{U.S.} \gtrless (\hat{1})_{G.B.}$$

depends upon

$$(27) \quad \left(\frac{P_A}{MPP_T}\right)_{U.S.} \gtrless \left(\frac{P_A}{MPP_T}\right)_{G.B.}$$

the relative magnitudes of the two components of equation (25). If for instance $(MPP_T)_{U.S.} \ggg (MPP_T)_{G.B.}$ then $(\hat{1})_{U.S.} > (\hat{1})_{G.B.}$ and the higher American land rental would tend to shift the graph of equation (22) toward the origin. The two shift parameters $\hat{1}$ and $(P_A/\overline{P})$ would then tend to place the American frontier inside the English one. The hypothetical conclusion being that America could be labor or capital abundant or both (if lower factor prices are taken as evidence of "abundance").

Since we would expect competition and labor mobility between agriculture and manufacturing to equalize the wage of unskilled labor, w_u, we should also expect that

$$(28) \quad (w_{u,A})_{U.S.} = (w_{u,M})_{U.S.}$$

i.e. the wage of unskilled labor should be the same in both sectors. However the factor-price-frontier for the manufacturing sector implies that

$$(29) \quad (\hat{w}_u)_{U.S.} > (\hat{w}_u)_{G.B.}$$

the wage rate for unskilled labor in America was higher than that for England, a result which is consistent with Allen's data. For the frontier derived from the agricultural production function we find on the basis of plausible assumptions about the relative magnitudes of P_A and MPP_T that

$$(29a) \qquad (\hat{w}_u)_{U.S.} < (\hat{w}_u)_{G.B.}$$

Now if we may speak of "the" real rage rate for unskilled labor in England as being some single value, then the American rate cannot be both above and below it. The two sector model employed does not give answers that are consistent with its own assumption, i.e., internal competition between the two sectors in each domestic economy.[16]

The formulation of the question of Anglo-American differences in technology with reference to capital-labor substitution in terms of the factor-price frontier model originally employed by Temin and re-specified by Fogel has thus come full circle with a return to the position that the question of factor scarcity is essentially an empirical one; and not one susceptible of treatment by known theoretical methods alone given the type of ordinal and conjectural estimates we are capable of making based on the state of our empirical knowledge about the first half of the nineteenth century.

Three final points on the above methodology are in order. First, Temin's paper connects the "real wage" with the abundance of scarcity of labor. In Figure 1.3 $w_2 > w_1$. If two economies faced the same supply curve of labor it would hardly be accurate to say

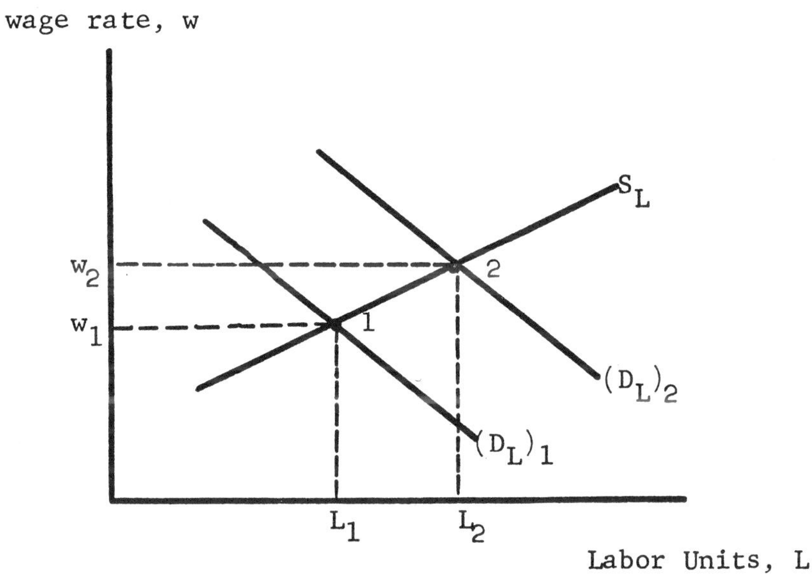

FIGURE 1.3

SUPPLY OF AND DEMAND FOR LABOR

that country 2 was labor "scarce" because $w_2 > w_1$, in both cases the labor market has cleared and both economies can get to points on their demand curves for labor. If however the higher wage rate in the second country resulted from an inelastic supply of labor as shown in Figure 1.4, given an identical demand curve

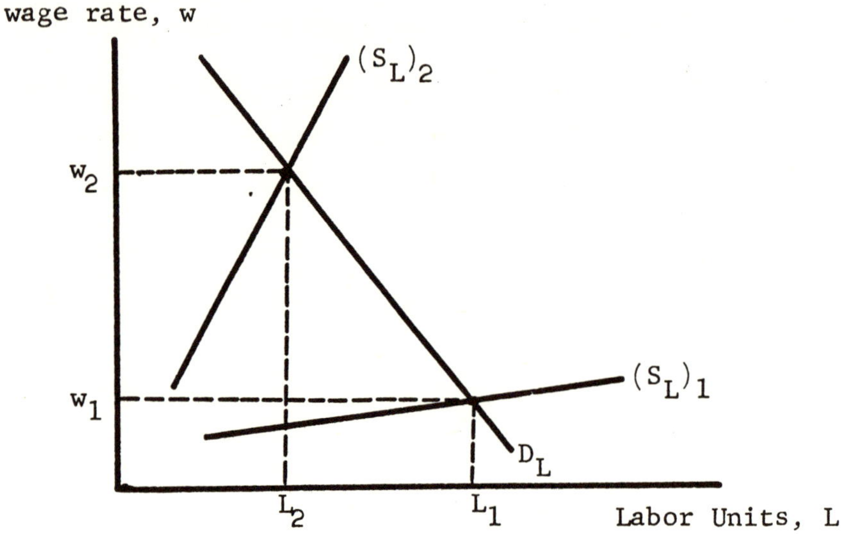

FIGURE 1.4

THE EFFECT OF DIFFERENT SUPPLY ELASTICITIES

for labor in the two countries then we could, with some justification, speak of "labor scarcity" in country 2, i.e. with a higher real wage less labor is called forth in country 2 even though labor markets clear in both countries. $(L_2 < L_1)$ Thus by itself the term "scarcity of labor" conveys very little information. A relevant concept in the discussion of the labor scarcity hypothesis as it relates to technological change is the elasticity of supply of labor, i.e., the responsiveness of the supply of labor services to changes in the wage rate. In this sense we may speak of "scarce labor," labor services not readily forthcoming as the wage rate is increased, an inelastic supply of labor. To date no discussion in the literature has approached this question of the labor scarcity hypothesis with an analysis of comparative conditions of labor supply. In this context the relative real wage in the U.S. and G.B. as used by Habakkuk, Temin and Fogel in order to speak of "scarce" or "abundant" labor has no bearing on the question of technological development.

The second point relates to the use of the factor-price-frontier itself to analyze the question of capital-labor substitution. Both Temin and Fogel implicitly are analyzing the question of why different factor-proportions prevailed in Britain and America. As Temin points out, assuming common technologies, it is relative factor prices that determine the optimal factor proportions. Nothing in the analysis of the factor-price-frontier can be used to determine prevailing relative prices. At best this methodology permits us to make controlled conjectures about the relative positions of the English and American frontiers.

Drummond has taken issue with Temin's use of a neo-classical production function in that its assumptions of continuously substitutable inputs, constant returns to scale, and economy-wide equilibrium were not appropriate to the early nineteenth century world of a limited range of techniques, i.e., feasible capital-labor combinations.[17] The problem, however, lies not so much with the restrictiveness employed on the form of the production function as with the lack of empirical information. It is this lack of data which prevents us from employing the factor-price-frontier to make bounded estimates of Anglo-American factor price differentials under any realistic specification of the manufacturing production function.

Throughout the Temin-Fogel analysis the assumption of common technologies is employed. However, this to a large extent assumes away the problem to which the labor scarcity hypothesis addresses itself, i.e., did different relative factor prices in England and America lead to technological divergence. Temin assumed the burden of explaining how different factor proportions could arise by virtue of different relative factor prices in two countries even if those countries employed a common technology.

The reason Temin did this was: "If a difference in technology is assumed, then the argument says nothing about its cause."[18] However, Temin never got around to the point that was presumed to be at issue, i.e., how do different levels of relative factor prices influence technological development and change. If America had a higher capital-labor ratio this fact alone tells us nothing about the nature of her technology and why it developed as it did.[19]

Finally, if as Temin and Fogel have assumed, the English and American economies have the same technologies, production coefficients, and sectoral compositions, how would it be possible for manufacturing prices to rise relative to the index of general commodity prices in one country and not the other? In other words is the fact that $(P_M/\overline{P})_{U.S.} > (P_M/\overline{P})_{G.B.}$ an implication of the model plus its assumptions? Clearly not because the model plus its assumptions imply two identical economies. In real terms the only way relative manufacturing prices can differ is if one economy has more abundant land than the other, this is how the production possibility curve for one country and its dual, with which we have been dealing, can shift outward relative to the other country. Different land endowments would also imply that the factor price frontiers for the two economies are not homothetic, as Fogel's diagrams indicate. The use of the shift parameter in this case constitutes a conflation of exogenous historical knowledge with the full set of implications of the model itself.

The Specification Dilemma

The Habakkuk-Temin-Fogel analysis then leaves the question of technological change in the American economy quite open. The principal, and by no means inconsequential, benefit deriving from this controversy has been a sharpening of the economic issues involved in the labor scarcity hypothesis as a potential explanatory device for understanding American technological development.

What has emerged in the analysis presented in this chapter is a conflict that many economic historians sense intuitively. The essential element in the proper specification of a model is that it includes variables and relationships that are just necessary and sufficient to properly define, on theoretical and logical grounds, the economic properties of some subset of economic experience. The economic theorist strives for general causes and explanations by emphasizing the relationships among a few key variables, and his models are correctly specified only when they measure up to that mark. The economic historian must deal with his subject not only with the criterion of descriptive reality in mind, but also with the additional requirement that he use economic theory properly.[20] It is no small task to be perfectly faithful to this dual mandate, and here the dilemma of specification in economic history arises. A theoretically correct specification of some model based on received economic theory, dealing with aggregative concepts, may be inappropriate for the subset of experience or class of phenomena that the economic historian has under consideration. At certain levels of analysis, for problems within the scope of economic history, there may well be a meaningful difference in what is correct specification from a theoretical or historical point of view.

This is to say that models, as received from economic theory, must be modified and re-specified so as to insure that they correspond closely with the range of experience or "facts" they purport to describe. In the cases analyzed in this chapter this re-specification led to the position where the information requirements of the model i.e., data on the cardinal values of the shift parameters, $(P_M/\overline{P})$, $(P_K/\overline{P})$, etc. exceeded the amount of information yielded by the model itself, $\hat{W}$.

The next point relates to a ground rule of legitimate specification. Temin's model, as specified by him, was sufficient to reject the dual scarcity position, while Fogel's re-specification of the manufacturing production function led to a re-instatement of the traditional position. A specification to be correct with regard to the rules of evidence on which scientific inquiry is founded must be theoretically and logically complete before the proposition at issue is brought to test. In other words, it is illegitimate to specify a model, a priori, in order to accept or reject a particular position, without fully testing and reporting the sensitivity of the conclusions drawn from the model to its specification.

All of this is not to detract from the enormous contributions of Professors Fogel, Habakkuk and Temin; nor is it meant as a disclaimer about the use of theory in economic history. It does suggest, however, that empirical problems are often best solved by empirical methods.

CHAPTER II

FACTOR SUBSTITUTION AND PRODUCTIVITY GROWTH RECONSIDERED

The implication of the Temin-Fogel models and of hypotheses of the labor scarcity variety is that the capital-labor ratio is some function of relative factor prices. Beyond this the use of smoothly continuous and differentiable economy-wide production function implies some rather large value of the elasticity of substitution. Under perfect competition this last condition would mean that the capital-labor ratio is responsive to changes in relative factor prices. The point being that optimal factor proportions will be uniquely and continuously determined by the relative prices, i.e., the high cost of labor relative to capital in America thus "explains" the relative capital intensity of American manufacturing.

Related to this point is the maintained hypothesis that technological change follows along, or results from a capital intensive production process. While models of the Habakkuk, Temin, Fogel variety never state this explicitly it is after all technological change which is presumed to explain most of the variation in efficiency between England and America. However, in models of this latter variety there is no causal connection between capital intensity, technological change, and productivity. That models of this type should add very little to our historical understanding of the process of technological change is not very surprising when we consider the above.

Factor Substitution When the Choice of Techniques is Limited

The first part of this section will consider the plausibility of the assumption of continuous economy-wide factor substitutability from a theoretical and empirical standpoint. The second part will develop an explanatory hypothesis concerning the connection between factor prices, capital intensity, and what is called technological change.

One approach to considering the possibility of substitution in economies in the early nineteenth century is to use knowledge derived from the history of technology in conjunction with some theoretical concepts about alternative productive activities in the Leontief system. The Leontief model can be generalized by relaxing the assumption of a unique correspondence between the input ratio and the level of output. This generalization can be carried out without affecting the "linear" character of technology. This is the Von Neumann Model which permits factor substitution by allowing output to be produced by a linear combination of adjacent activities, i.e., points of fixed input ratios.[1] Given, say, three production techniques, each of which is of the fixed coefficient type, we can obtain a linear combination of these activities which corresponds to a portion of an isoquant over which substitution is possible.

Figure 2.1 shows three production techniques of the fixed-input ratio or Leontief type. The input combinations (K_1, L_1), (K_2, L_2) and (K_3, L_3) each represent the same level of physical output. As shown in Figure 1.2 technique 3 would not be chosen because it is clearly inferior to techniques 1 and 2 . It is inferior to technique 1 because $K_3 = K_1$ but $L_3 > L_1$; and 3 is inferior to 2 because $L_2 = L_3$ but $K_3 > K_2$, i.e., technique 3 uses more of at least one input and not less of the other input than each of the other techniques. At any point on the sloping portion of the combined isoquant represented by the dark line connecting

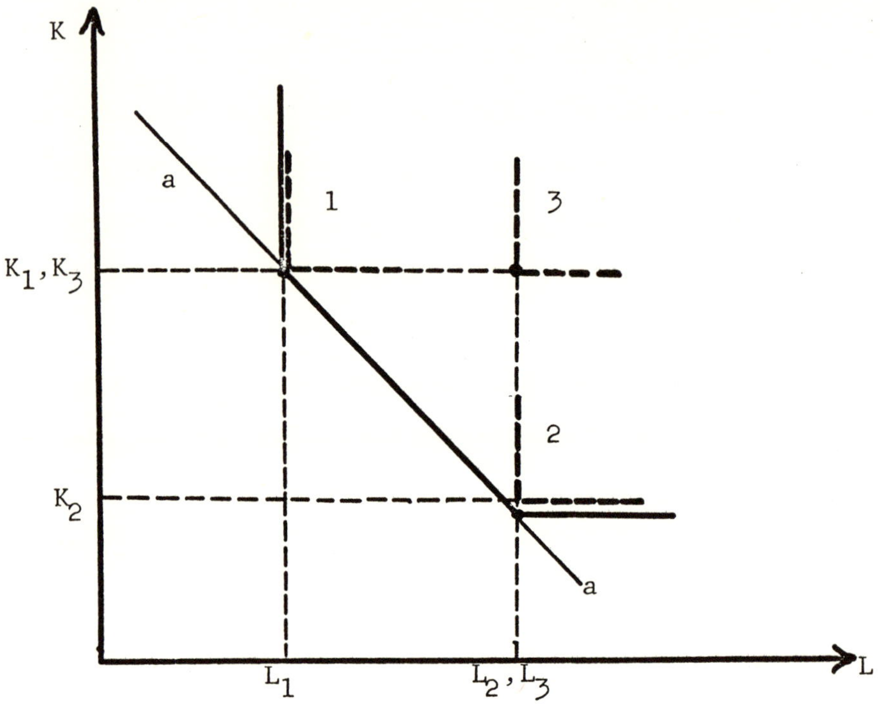

FIGURE 2.1

THE VON NEUMANN MODEL

points 1 and 2 processes 1 and 2 are both in use simultaneously. This sloping portion of the curve could represent different proportions of the output of the arms industry made by gunsmith methods and the Armory or factory system, assuming these were the two known techniques for producing firearms and that the optimal capital-labor ratio for each technique was constant. If three or more techniques had been available then the isoquant would have more "breaks" or discontinuities. The appeal of this formulation is that, in terms of the number of techniques available to an early nineteenth century industry, there was only a sharply limited number of technical alternatives. In this kind of a world factor substitutability arises only as a "linear" possibility, i.e., from the combinations of different proportions of output being produced by alternative adjacent technologies. This is so even though the techniques themselves do not permit factor substitution because they are of the fixed input ratio type.

An interesting implication of the Von Neumann Model is the insensitivity of the capital-labor ratio to relative factor prices. For example, in Figure 2.1, a-a is the budget line showing relative factor prices. Given this set of relative prices, production could be carried out at point 1 or point 2 or at any point between these two techniques. In fact the relative price w/r could go from zero at point 2 to infinity at point 1 before bringing about a shift in technique from 1 to 2. What is more important is that given the price line a-a we cannot tell what factor proportion will prevail for an industry, as a logical possibility w/r could go from 0^+ to $+\infty$ and the industry could continue to employ the same overall capital labor ratio. In other words the elasticity of substitution of labor for capital would be zero for the industry as a whole. Thus we could in principle derive a production function that had all the essential properties specified by Temin; but would be one for which relative prices would have no effect on the chosen capital-labor ratios. While it may be argued that production functions of the Von Neumann type may be valid for an industry, when we sum these functions over all industries to get a production function for an economy, this aggregate function will possess only the minutest discontinuities and be, for all practical purposes, smoothly continuous. Thus it would be argued economy-wide capital-labor ratios would change responsively to small shifts in relative prices. While indeed the isoquants and price lines are capable of being aggregated, the behavior underlying each industry's (or firm's) choice of factor proportion is not. That is if as a behavioral proposition firms and industries have negligible elasticities of substitution of labor for capital with respect to relative prices what then does it mean to have a large, aggregate elasticity of substitution? As a behavioral proposition . . . precisely nothing. In the nineteenth century American economy the decision to substitute one factor for another is made at the level of the firm (and in some cases the industry), not by some central planner. Then if we seek to understand why one economy is more, say, capital intensive than another we must look at pressures and influences which impinge at the level of analysis which is appropriate to the phenomenon we are trying to understand. The notion that relative factor prices determine optimal factor proportions is essentially a proposition about behavior at the level of the firm. The implication of the Von Neumann Model is that changing relative factor prices do not necessarily bring about changes in factor proportions when the number of technical alternatives is small. In the world during the first half of the nineteenth century, the number of alternative techniques open to any industry was limited and the Von Neumann Model parallels this reality fairly closely. In understanding how technological change comes about we are to some extent concerned with behavior, i.e., how a given technique is chosen over some alternative. The decision to incorporate the technology embodied in a piece of machinery is the result of purposeful, deliberative behavior on the part of some individual, firm or industry. This aspect of technology, how it is developed and implemented, is then most fruitfully studied by examining the pressures and conditions attending its utilization *at the appropriate unit of analysis*. At the level of the firm, where this decision-making process comes into play, relative prices are not the crucial consideration when the choice of techniques is limited.

Factor Substitution and Labor Productivity Growth

Having argued above that the appropriate level of analysis of factor substitution is at the level of analysis where decision-making takes place, we will now turn to a broader analysis of this mechanism. The purpose of the following aggregate analysis is to provide a theoretical and empirical background for viewing the development of the manufacturing sector in the American economy in the nineteenth century. Following a method of analysis employed by Salter, some hypothetical values for changes in the share of capital given a range of relative price changes will be developed.[2] Using these changes in relative shares the percentage increases in labor productivity attributable to capital-labor substitution could be calculated. These last mentioned increases in labor productivity will then be compared to the measured or observed increase in labor productivity in the nineteenth century to determine what proportion of the increase in total labor productivity might be attributed to the substitution mechanism and hence the importance of factor substitution as an explanatory factor in the growth of labor productivity or what might be termed technical efficiency.

As has been pointed out above the influence of differential, relative prices on factor proportions and hence factor shares is governed by the elasticity of substitution of labor for capital. By allowing a range of possible values for the elasticity of substitution and the initial shares of capital we can estimate the percentage change in capital's share for various changes in relative factor prices.[3]

In calculating Table 2.1 the sign of dr/r is positive, that is the term (w/k) is taken to increase as wages rise relative to capital costs. As Table 2.1 shows relative shares are insensitive to rather large shifts in relative factor prices for a range of feasible elasticities of substitution. The 30% change in relative factor prices corresponds roughly to the percentage differential of American labor costs over and above English labor costs spoken of in the literature.[4] Of course since we are not certain of absolute differentials in capital costs between England and America we cannot be certain that the 30% to 90% change in relative factor prices could correspond to the relative factor price differential between the two countries in the early nineteenth century. However, if capital costs were absolutely higher in the United States then the use of the 30% change in relative prices in Table 2.1 considerably overstates the differential percentage in relative factor prices between the two countries, and consequently overestimates the differences in factor shares assuming common technologies. While recognizing the tenuous basis of these calculations we could use the hypothetical estimates of Table 2.1 as a range of values for changes in factor shares to estimate the proportion of the increase in total labor productivity attributable to substitution. Fortunately, estimates of labor's share for the last five decades of the nineteenth century are available and we shall use these observations later on. Since we are assuming that product exhaustion holds in this theoretical world, the change in labor's share is simply equal and opposition in sign to the calculated change in capital's share. Using the range of estimates in Table 2.1 the percentage change in labor's share varies from

TABLE 2.1. THE CHANGES IN CAPITAL'S SHARE IN NET OUTPUT CAUSED BY FACTOR SUBSTITUTION

Initial share of capital (%)	For 30% increase in relative factor prices					For 60% increase in relative factor prices					For 90% increase in relative factor prices				
	σ					σ					σ				
	0	.5	1.0	1.5	2.0	0	.5	1.0	1.5	2.0	0	.5	1.0	1.5	2.0
	% change in share of capital					% change in share of capital					% change in share of capital				
20	-3.7	-1.8	--	+1.8	+3.7	-6.0	-3.0	--	+3.0	+6.0	-7.6	-3.8	--	+3.8	+7.6
30	-4.8	-2.4	--	+2.4	+4.8	-7.9	-3.9	--	+3.9	+7.9	-10.0	-5.0	--	+5.0	+10.0
40	-5.5	-2.8	--	+2.8	+5.5	-9.0	-4.5	--	+4.5	+0.0	-11.4	-5.7	--	+5.7	+11.4
50	-5.8	-2.9		+2.9	+5.8	-9.4	-4.7		+4.7	+9.4	-11.8	-5.9		+5.9	+11.8

approximately -12% to +12% for the assumed values of initial share of capital, elasticity of substitution, and percentage change in relative prices.

Table 2.2 shows the expected change in labor productivity that would result from factor substitution above, given the observed changes in wage' share over the last five decades of the nineteenth century.[5] As opposed to the hypothetical change of ±12% in labor's share developed in Table 2.1, the range estimated by Edward Budd, ±7+, is used in Table 2.2.

Table 2.2 indicates that for the observed percentage changes in labor's share of net income in the industrial sector and the assumed range of increase in relative factor prices that average, decadal increases in labor productivity of from 15.8 to 26.0% could have resulted from substitution alone. (Though not shown here the corresponding figures developed from the ±12% change in labor's share of Table 2.1 would be 14.3 to 19.8%. These hypothetical estimates are obtained by assuming the four separate values of capital's share shown in Table 2.1 for each value of the percentage change in labor's share for each of the three relative price changes.)

TABLE 2.2. PERCENTAGE INCREASE IN LABOR PRODUCTIVITY DUE TO FACTOR SUBSTITUTION IMPLIED BY THE OBSERVED VALUES OF PERCENTAGE CHANGE IN LABOR'S SHARE IN INDUSTRY'S OBSERVED INITIAL SHARE OF INCOME, AND ASSUMED CHANGE IN RELATIVE FACTOR PRICES

Year	Observed change in service income's share of total income in industry	Values of σ implied by the observed changes in labor's share when the increase in relative prices is			% change in labor productivity implied by these elasticities when the increase in relative prices is		
	(%)	30%	60%	90%	30%	60%	90%
1	2 (a)	3	4	5	6	7	8
1850-1860	-0.8	1.14	1.09	1.07	9.0	15.5	20.0
1860-1870	+5.8	2.02	1.62	1.49	21.8	28.2	32.8
1870-1880	+0.6	1.11	1.07	1.05	10.4	16.3	20.2
1880-1890	+7.0	2.26	1.78	1.62	20.8	26.6	30.6
1890-1900	-7.2	2.41	1.88	1.69	17.2	23.2	26.5
Decadal average % increase in labor productivity					15.8	21.0	26.0

(a) The figures are derived from the estimates of service income as a percentage of total income, by sector, by component in Edward C. Budd, "Factor Shares 1850-1910," in Trends in the American Economy in the 19th Century, Studies in Income and Wealth, Vol. 24, NBER, Princeton University Press, 1960, Table 7, page 382. Budd's concept of service income is intended as a measure of labor's share. The % changes used in this table are for the industrial sector.

Given these expected percentage increases in labor productivity due to substitution of capital for labor, we will now examine the aggregate changes in labor productivity in manufacturing for the American economy during the nineteenth century to determine what proportion of the total increase in labor productivity could be accounted for by factor substitution. Table 2.3 represents a pooling of data developed separately by Robert Gallman and Stanley Lebergott on value added in manufacturing, commodity output, and labor force estimates at ten year intervals starting in 1839. The measure of productivity employed in Table 2.3 is output per unit of labor, (Q/L).

From Table 2.3 the average decadal increase in labor productivity is 22.1% in manufacturing and 13.4% for total commodity output as defined by Gallman. Thus is relative factor prices (w/k) were increasing by 30% each decade in the American economy, substitution alone would account for about 70% of the total observed increase in labor productivity in the manufacturing sector. When the percentage change in labor productivity in manufacturing over the entire 50-year period is considered, the maximum proportion of the increase that can be accounted for by substitution is only 10%. This occurs

TABLE 2.3. VALUE ADDED BY MANUFACTURE, TOTAL COMMODITY OUTPUT, MANUFACTURING LABOR FORCE, OUTPUT PER UNIT OF LABOR AND % CHANGES IN OUTPUT PER UNIT OF LABOR

Year	V.A. by mfgrs. in 1879 prices (millions)	Total com. output in 1879 prices (millions)	Mfg. labor force (thousands)	Total labor force (thousands)	Output per unit of labor, mfg.	Output per unit of labor, total	% change in product. of labor by decade, mfg.	% change in product. of labor by decade, total
	1	2	3	4	5 = 1 ÷ 3	6 = 2 ÷ 4	7	8
1839	$ 190	$ 1,094	500	5,660	$ 380	$193		
1849	488	1,657	1,200	8,250	407	201	+7.1	+4.1
1859	859	2,686	1,530	11,110	560	242	+37.8	+20.4
1869	1,078	3,271	2,470	12,930	437	253	-22.0	+4.5
1879	1,962	5,304	3,290	17,390	596	305	+36.5	+20.5
1889	4,156	8,659	4,390	23,320	950	372	+59.5	+22.0
1899	6,252	11,751	5,895	29,070	1,063	405	+11.9	+8.9

	Mfg.	Total
a) Average decadal increase in labor productivity	22.1%	13.4%
b) Increase in labor productivity over 60 year period	279.0%	210.0%
c) Increase in labor productivity, 1849-1899	262.0%	200.8%

Notes to Table 2.3 on page 47.

Notes to Table 2.3:

Columns 1, 2

Robert E. Gallman, "Commodity Output, 1839-1899," in _Trends in the American Economy in the 19th Century_, Studies in Income and Wealth, Vol. 24, NBER, Princeton University Press, 1960. Value added by manufacturer in 1879 prices, Table A-5, page 56. Commodity output Table A-1, variant A, page 43.

Columns 3, 4

Stanley Lebergott, "Labor Force and Employment, 1800-1960," in _Output, Employment and Productivity in the U. S. after 1800_, Studies in Income and Wealth, Vol. 30, NBER, 1966. Figures on manufacturing labor force and total labor force are from Table 1, page 118. Columns 3 and 4 are Lebergott's labor force estimates for 1840, 1850, etc., therefore the output per unit of labor figures do not represent output and labor force estimates for exactly the same year.

in Table 2.2 if we assume that relative prices increased 90% over the 50-year period from 1850-1900.

While recognizing the hypothetical nature of the calculated increases in labor productivity due to factor substitution and the crude nature of the comparisons made, it is still evident that factor substitution itself can account for at most but a small portion of the measured increase in labor productivity during the last five decades of the nineteenth century in the American economy. To turn this argument around if the American and English economies were alike in every respect including technology, as Temin and Fogel have assumed, with the exception that relative factor prices increased 90% in the American economy during the last five decades of the nineteenth century, then substitution would account for at most one-tenth of the observed increase in labor productivity in manufacturing in America. As Temin has shown the labor scarcity hypothesis is really a simple proposition, i.e., the relationship between relative factor prices and factor proportions. As such, this hypothesis does little to further our understanding of the link between capital intensive production and technological progress. Habakkuk has attempted to forge such a linkage on theoretical and conjectural grounds.[6] However, as has been shown the factor substitution mechanism alone, replacing homogeneous labor with homogeneous capital of an unchanged quality, does not account for a very significant portion of the observed increase in labor efficiency in the nineteenth century American economy.[7] The crucial point to note is that the labor scarcity hypothesis is not necessarily an economic hypothesis. Habakkuk can connect capital intensity and technical advance only by appealing to a rather conjectural theory of invention, i.e., "technical possibilities were richest at the capital-intensive end of the spectrum." Thus if the labor scarcity hypothesis is to have any validity in explaining technical advance it has to be bolstered by additional arguments, over and above factor substitution, many of which are non-economic in nature.

Empirical Evidence on Capital-Labor Substitution in the Nineteenth Century American Economy

The timing and magnitude of capital-labor substitution in the manufacturing sector of the American economy can be examined more directly. As a starting point for this analysis the capital value-added ratios given by Simon Kuznets will be modified.[8]

The capital-value added ratios start in 1880 and are measured in volumes based on 1929 prices. They are as follows:

Year	Ratio of total capital in mfg. to V. A. by mfg. in 1929 prices
1880	1.51
1890	1.65
1900	1.88

Since these ratios are in 1929 prices and they will later be used with Gallman's estimates of value-added in manufacturing in 1879 prices, it is necessary to deflate both numerator and denominator in the Kuznets estimates to conform to ratios in 1879 prices. An example of the operation to be performed, for Kuznet's 1890 estimate is:

$$\frac{\text{Total capital in mfg. in 1879 prices}}{\text{V.A. in mfg. in 1879 prices}}$$

$$=$$

$$\frac{\dfrac{\text{Total capital in mfg. in 1890 prices}}{\text{1929 price index of mfg. capital}}}{\dfrac{\text{V.A. by mfgs. in 1890 prices}}{\text{1929 price index of V.A. by mfgs.}}}$$

$$\times$$

$$\frac{\dfrac{\text{1929 price index of mfg. capital}}{\text{1879 price index of mfg. capital}}}{\dfrac{\text{1929 price index of V.A. by mfgrs.}}{\text{1879 price index of V.A. by mfgrs.}}}$$

The deflators for manufacturing capital and value-added were obtained by using the implicit deflators in the Kuznets data. These indices give rise to estimates of the ratio of total capital to value-added by manufactures based on volumes in 1879 prices. The new estimates are presented in Table 2.4.

The three ratios of total capital in manufacturing to value-added in manufacturing in 1879 prices imply a 1-1/8% rate of growth in the capital-net output ratio in the manufacturing sector in the American economy in the last two decades of the nineteenth century. This trend rate would give a capital-output ratio of unity in 1839, assuming that capital deepening was progressing at a uniform rate over the last six decades of the century. Column 2 of Table 2.5 shows the value of the capital-value added ratio in manufacturing resulting from a backward extrapolation of the 1-1/8% rate of growth. Multiplying these ratios in Column 2 by Gallman's estimates of value-added in manufacturing Column 3 gives total capital in manufacturing, Column 4, for the period 1839-1899 at ten-year intervals. Dividing the total capital estimate by Lebergott's labor force figures, Column 5, yields an estimate of capital per worker in 1879 prices. Column 7 of Table 2.5

TABLE 2.4. ESTIMATES OF TOTAL CAPITAL IN MANUFACTURING TO VALUE-ADDED IN MANUFACTURING BASED ON VOLUMES IN 1879 PRICES

Year	Total capital in mfg. to V.A. based on 1929 prices	Estimated deflator, capital	Estimated deflator, V.A.	Total capital in mfg. to V.A. based on 1879 prices
	1	2	3	4 = 1 x (2÷3)
1880	1.51			1.56
1890	1.65	1.74[a]	1.68[b]	1.71
1900	1.88			1.95

Notes:

(a) This deflator is obtained from S. Kuznets, <u>Capital in the American Economy</u>, Table R-33, page 596 by dividing Kuznet's estimate for 1879 of the dollar value of net producer's durables in 1929 prices by the estimate of this magnitude in current (1879) prices.

(b) This deflator is obtained from S. Kuznets, <u>Capital in the American Economy</u>, Table B-2, page 511, col. 1, it is the implicit index in the flow of goods to consumers (variant I) for the period 1879-1882 divided into the base index for 1929 = 100.

TABLE 2.5. TOTAL CAPITAL TO VALUE-ADDED IN MANUFACTURING, VALUE-ADDED, TOTAL CAPITAL IN MANUFACTURING, MANUFACTURING LABOR FORCE, TOTAL CAPITAL TO LABOR FORCE IN MANUFACTURING IN 1879 PRICES

Year	Total capital in mfg. to V.A. (billions)	V.A. in mfg. (billions)	Total capital in mfg. (billions)	Mfg. labor force (millions)	Total capital in mfg. to mfg. labor force ($)	% change in capital-labor ratio in mfg.
1	2	3	4 = 2 x 3	5	6 = 4 ÷ 5	7
1839	1.00	.190[a]	.190	.500[b]	380	
1849	1.06	.488	.518	1.200	432	+13.7
1859	1.25	.859	1.072	1.530	700	+62.0
1869	1.40	1.078	1.510	2.470	612	-12.6
1879	1.56	1.962	2.998	3.290	911	+49.0
1889	1.71	4.156	7.120	4.390	1625	+78.3
1899	1.95	6.252	12.170	5.895	2065	+27.1

(a) Robert E. Gallman, "Commodity Output," op. cit.

(b) Stanley Lebergott, "Labor Force and Employment," op. cit.

shows that a sustained deepening of capital per worker took place in the economy between 1849-1889, with the exception of the decade spanning the Civil War years. Given the straight line extrapolation of the growth in the capital-output ratio, it is possible that these results, indicating that substitution of capital for labor did not take place to a significant extent in the first half of the nineteenth century, placing the date of the upswing in the capital-labor ratio too early in the century. An exponential growth of the capital-output ratio would have led to smaller, more closely spaced capital-labor ratio estimates in the early decades of the 1839-1899 interval and thus diminish the level of percentage change in the capital-labor ratio in this initial part of this period.

Another measure of the extent of capital-labor substitution may be found by using single year estimates of gross capital formation in manufacturing durables. Column 6 in Table 2.6 represents the annual, new, gross investment per worker and this captures to some extent the magnitude of the substitution actually taking place from year to year as a result of conscious decision-making by entrepreneurs. Table 2.6 shows that a sharp rise in annual gross investment per worker occurred in the interval 1869-78. The figure of $179

TABLE 2.6. SINGLE-YEAR ESTIMATES OF GROSS CAPITAL FORMATION IN MANUFACTURING DURABLES IN 1860 PRICES, MANUFACTURING LABOR FORCE, AND ANNUAL GROSS CAPITAL FORMATION PER WORKER

Gallman's over-lapping decades	Mfg. durables (billions)	Center-year for mfg. durables single-year est.	Aggregated single-year est. mfg. durables (billions)	Mfg. labor force (millions)	Annual gross capital formation per worker ($)
1	2	3	4	5	6 = 4 ÷ 5
1834-1843 1839-1848	.032 .045	1840	.039	.50	77
1844-1853 1849-1858	.074 .113	1850	.094	1.20	78
1869-1878	.441	1860 1870	(.277)[a] .441[b]	(1.53) 2.47	(181) 179
1874-1883 1879-1888	.685 1.08	1880	.883	3.29	268
1884-1893 1889-1898	1.47 1.80	1890	1.64	4.39	374
1894-1903	2.30	1900	2.30	5.895	390

Notes to Table 2.6 on page 56.

Notes to Table 2.6:

Columns 1 , 2

Robert E. Gallman, "Gross National Product in the U. S., 1834-1909," in Output Employment and Productivity in the U. S. After 1800, Studies in Income and Wealth, Vol. 30, NBER, 1966, p. 34, Table A-3, Column 1.

Columns 3 , 4 , a, b

The center-year averages of Gallman's data were obtained by averaging the single year estimates for the over-lapping periods. The estimate of .277 billion for 1860 is the simple average of .113 for 1849-1858 and .441 for 1869-1878. The .441 billion for 1870 is Gallman's unadjusted single year estimate for 1869-1878.

Column 5

Stanley Lebergott, "Labor Force and Employment," op. cit.

per worker in prices of 1860 is obtained by dividing Gallman's single year estimates for this period by Lebergott's estimate of the labor force in manufacturing for 1870. The comparable figure for the 1849-58 interval is $94 per worker obtained by dividing the single year estimates by the manufacturing labor force in 1850. On this basis the earliest date for a sustained rise in gross investment per worker, indicating substitution of capital for labor, was 1869 or 1870. If the single year estimates for these intervals are averaged a single year estimate of gross capital formation in manufacturing durables of .277 billion is obtained centered on 1860 at the earliest. This figure, .277 billion, divided by the 1860 manufacturing labor force, yields an estimated gross investment per worker of $181 in 1860, a doubling of gross investment per worker over the comparable figure for 1850. Thus by pushing these estimates to arrive at the earliest date for a sustained upturn in gross investment per worker it might be said this increase occurred by 1860. However, given the estimated decrease of 12.6% in the capital-labor ratio in manufacturing between 1859-1869 shown in Table 2.5, the large sustained substitution of capital for labor, assumed to be characteristic of the American economy throughout the nineteenth century, most likely did not occur until after the termination of the Civil War.

This chapter has presented theoretical and empirical evidence on the role played by capital-labor substitution in the American economy during the last six decades of the nineteenth century. The model employed showed that the maximum proportion of the observed increase in labor productivity that could be attributed to factor substitution in the period is about 10%. The empirical material drawn from the work of Gallman, Kuznets and Lebergott, while providing only approximate measures of the concepts employed, indicates that sustained increases in the substitution of capital for labor did not occur until 1850 at the earliest and possibly as late as the mid-1860's. This is not to say that factor substitution did not take place at all in the first half of the nineteenth century, indeed, some substitution of capital for labor was going on. However, the magnitude of this process was wholly different between the two halves of the century. In particular it is this last point which should cause us to suspect the general hypothesis that factor substitution occasioned by the high relative price of labor was a dominant characteristic of American manufacturing throughout the nineteenth century. The frontier of our empirical knowledge on factor substitution is reached in the beginning of the fourth decade of the century when Gallman's estimates of value-added in manufacturing start. If, however, as the above evidence leads us to suspect, factor substitution was a limited phenomenon in the manufacturing sector in the first half of the nineteenth century its role in any explanatory hypothesis must be likewise diminished. It is precisely this period, the first half of the nineteenth century, in which American technology developed its special features, and it is this development which needs a more thorough exploration and explanation.

CHAPTER III

AN ESTIMATE OF SOCIAL SAVINGS AND HUMAN CAPITAL ATTRIBUTABLE TO IMMIGRATION: 1790-1860

The Conceptual Framework

So far the discussion centered around the controversies arising out of the labor-scarcity hypothesis has concentrated on the substitutability of physical capital and labor. This way of viewing the problem of the divergence in technological development between the United States and Great Britain has tended to ignore the magnitude of two effects, operating with varying strengths in the pre-Civil War American economy. The first is the social saving accruing to the American economy as a result of having that portion of its productive population added by net immigration each year being raised to productive age by the resources of an external, national economy, e.g., England, Ireland, Germany, etc. The second effect, quite distinct conceptually, is that each immigrant entering the labor force, by virtue of such skills as he might possess, represented a transfer of human capital. Each of these concepts has implications for the character of technical development, and the magnitude of these effects, which will be presented in this section, should serve to alter conventional interpretations of the importance of physical capital for economic growth in the nineteenth century.

Turning to the question of the social saving in the United States economy in the period 1790-1860 the limitations of and assumptions embodied in the estimate must be carefully stated. The concept of "social saving" has been employed by Professor Fogel in his study of the contribution of the railroads to American economic growth.[1] The term social saving as applied here is the social resource cost of raising a human being to productive age at the level of subsistence. The productive age is taken as that age at which an individual by the application of unskilled labor can produce a social product equal to his consumption requirements at or near the subsistence level. The point here is that to raise infants to the productive age requires resources. If productive individuals migrate to an economy, that economy receiving the flow has saved the social cost of raising that portion of its labor force to the productive age. These social savings represent a bundle of resources that might be employed in other alternatives. The hypothetical alternatives in the extreme would be either to increase consumption per capita or to invest these resources. Both of these alternatives require various assumptions about the distribution of capitalists' and workers' MPC's, time preferences, rates of return and the like. The purpose of the estimate is to show the quantity of resources that were available for re-channeling to, say, physical capital formation in the United States. This is not to say that all of the social savings attributable to immigration were directed toward the formation of physical capital. The point here is to indicate that these resources were available for physical investment _should the economy have so chosen_.

One of the most explicit statements of the magnitude of the social saving resulting from immigration to the U. S. is the following:

> Attention may be drawn to the interesting statistical computation of Corrado Gini that even the enormous capital stock characteristic of the United States economy is not more than the equivalent of the burden in consumption goods and in such services as health, education, and other provisions for the immigrants--a burden which the United States was enabled to save by shifting it to the European mother countries of the immigrants.[2]

A milder version of this statement appears in the work of Professor Thomas Cochran.

> Immigration to a nation essentially short of labor was unquestionably a stimulant to economic growth. Another country had paid for the immigrant's unproductive youthful years, and he came to the United States ready to contribute his labor at a low cost.[3]

In either case statements such as the above tend to leave the impression that there was a burden of considerable magnitude, in Singer's words "equivalent to the U. S. physical capital stock," transferred by immigration from America to Europe; and that this saving in resources was an important determinant of growth in the nineteenth century.

Social Saving Inflows to the American Economy: 1790-1860

At the outset it should be stated that the following estimate of social saving is an order of magnitude estimate only, resting quite crucially on some major assumptions of concept and fact. The following calculation is based in part on a set of variables that are demographic in nature. Such magnitudes as infant and child mortality and age distribution of the migrant stream play an important role in the estimates.

Column 1 of Table 3.3 shows the annual number of immigrants, male and female, of all ages coming to the United States from 1790 to 1860. These annual totals exclude returning citizens. The data for the period 1790-1804 are taken from Blodgett's, Economica, since immigration figures prior to 1820 are not given in the Historical Statistics of the United States.[4] Blodgett's figures include freemen and slaves. The figures are at considerable variance with other sources for this period. The Bureau of the Census has estimated that from 1791-1808 there were 70,000 slaves imported into America.[5] Blodgett shows a total of 65,000 freemen and slaves for the period 1790-1801, which in addition to the estimates of 1,500 per annum for the period 1805-1808 brings total immigration to 71,000 over the period comparable to the Census estimate. If both sets of data are correct then no freemen migrated to America between 1790 and 1808, an improbable event. The Census estimates correspond closely to estimates of slave importations in the eighteenth century made by J. Potter.[6] The following quotation from Rubin further clouds the issue of the composition of the migrant flow in this early period.[7]

> For the years 1790 through 1820 a series of immigration data on an annual basis is not

> available. According to early writers on immigration, it is estimated that 250,000 immigrants came to the United States between 1783, the close of the Revolutionary War, and 1820 when immigration statistics were first collected and published on an annual basis. It is my impression that this figure did not include the non-white population, primarily slaves, who were legally imported from Africa until 1808 and for many years thereafter smuggled into the country.

Rubin's estimate of 250,000 arrivals between 1783 and 1820 corresponds closely with the figure for this period given by Bromwell.[8] Since the Census figures after 1820 did not include slave importation, and since Bromwell treats the figures before and after 1820 as additive, we can make a weak inference that Bromwell's estimate of 250,000 immigrants from 1783 to 1820 was exclusive of slaves. After prohibition of the slave trade in 1808 smuggling continued. In 1818 a Congressional Act increasing rewards to informers checked this traffic. Finally, in 1820 Congress declared the maritime slave trade to be an act of piracy and though smuggling diminished it never completely ceased. The official Census position is that after 1820 the importation of slaves was "inconsiderable." The best evidence available to date comes from the work of Herbert Heaton who studied the industrial immigration from Great Britain in the period 1783-1812.[9] Heaton's work rests solidly on his use of primary data in the U. S. National Archives, primarily State Department records. Heaton found that,

> When business was booming on both sides of the Atlantic, migration rose to a high tide and at least ten thousand Britons landed on American soil in a good year . . . There were more good years than bad ones in the United States between 1783 and 1812 . . .

The weight of available evidence seems to resolve the dilemma posed by the contrary series in Blodgett and the Census estimates in favor of the view that there was a substantial migration of freemen between 1790 and 1815. The work of Rubin and Bromwell indicate 250,000 non-slave immigrants arrived from 1783-1820.

The gross immigration figures in Column 1 from 1790-1804 are Blodgett's figures on immigration which in light of the above, gave a fair portrayal of the level of non-slave immigration.[10] Throughout the computations of social savings and human capital inputs to the U. S. economy the contribution of slaves will be omitted This is not to say that there were no social savings due to slave holding nor that the human capital embodied in the slave population was inconsiderable. Rather it reflects the incomplete state of our knowledge relative to acquisition of slaves from their native lands. If the purchase price (command over real resources) was transferred from an American planter to a trader who purchased U. S. or foreign commodities for investment in the U. S. economy, then there was a social saving attributable to the slave represented by the transfer of these resources within the same economy. However, if a slave master used these funds to export American resources abroad for investment or consumption, then the resources represented by the purchase price were lost to the U. S. economy and there was no social saving. Thus whether there were social savings arising out of the use of slaves depends on the disposition of funds, acquired from sales to American planters, in the hands of slave traders. Due to the lack of evidence on this point, and considering that slave importations between 1791 and 1808 were approximately 70,000, and not growing substantially after that date; the impact of the slave trade will be ignored in the subsequent calculations.

The period from 1804-1815 was dominated by the hostilities between Britain and France. Bromwell has traced the effects of the various decrees and blockades from 1806-1815 and has concluded that, "Thus, from 1806, was the stream of migration pent up at its fountain."[11]

In the Census statistics for 1860 estimates for immigration from Great Britain and Ireland are given for 1815, 1816, 1817, 1818, and 1819 as 1,209, 9,022, 10,280, 12,429, and 10,674, respectively.[12] Thus from 1805-1815 an estimated annual flow of 1,500 immigrants to the U. S. is used. This figure is largely guesswork and rests mainly on the 1,209 figure given for 1815 as the immigration from Great Britain and Ireland alone. For 1817 Bromwell's estimate of "not less than 22,240" is used.[13] Since the number of returning Americans in 1817 was probably large the net migration for this year was taken as 75% of Bromwell's gross estimate. In the decade 1820-30, inclusive, the average percentage of migration estimates excluding returning citizens to those estimates including returners is 80%.[14] Thus the estimate of net immigration for 1817 is approximately 16,500. This figure is not greatly at variance with the immigration of 12,429 for the succeeding year from Great Britain and Ireland alone.

From 1820 to 1860 the figures on immigration to the U. S. are obtained from <u>Historical Statistics</u>, Table C88-114 which excludes returning citizens. In subsequent calculations those immigration estimates before 1820 are multipled by 80% to arrive at an estimate of net migration in this earlier period. Walter Willcox, whom Kuznets has described as "the dean of American demographers," has estimated the percentage that net migration was of gross as 100% for the period 1820-1830.[15] Thus adjusting the figures in the 1790-1820 period by an 80% ratio of net migration to gross should bias the result downward.

Columns 2, 3 and 4 are the annual percentage distributions of immigrants in the three age groups: under 15 years, 15-40 years, and over 40. Those arrivals for which no age was reported were included in the under 15 years category. This procedure again results in an understatement of the social savings, particularly for 1830, 1850, 1856 and 1857.

Columns 5, 6 and 7 show the numbers of net migrants in each of the three age brackets. Before 1820 it is assumed that return migration affected all the age groups uniformly. This assumption gives rise to a slight upward bias in the social savings calculation.

Columns 8 and 9 represent the resultant of a set of assumptions concerning age distribution in the under 15 years category, infant mortality, life expectancies in the early years, nursery costs of infants, the social subsistence cost of raising population to productive age, and finally the productive age itself.

A uniform distribution of ages is assumed in the under 15 years bracket.[16] Thus the average age of an immigrant in this category is taken to be 7.5 years. Other symmetric distributions would also give rise to an expected age of 7.5 years, e.g., the mean of a normal distribution with no skewness.

Table 3.1 shows the number of deaths per 10,000 in the 5-year age groups between 0 and 20 years for selected districts. Since these figures are from 1860 life tables the following mortality calculations tend to bias downward the annual estimates of social savings in the earlier years given that the probabilities of life were improving over the period 1790-1860.

The average number of deaths per 10,000 entering the 0-10 age group over the four districts is 4,849 with 5,151 surviving to the next bracket, 10-15. By interpolation the average number of deaths in the 0-7.5 year age bracket is 4,634 with 5,366 surviving to the next group.

For those who die in the 0-7.5 year interval the assumed number of years of life is 18 months. The _8th Census_ gives the average age of those dying under 20 as: 4.18 years, District I; 4.01 years, District III; 3.95 years, District V; and 3.86 years, District VII.[17]

TABLE 3.1. NUMBER OF DEATHS PER 10,000 IN VARIOUS DISTRICTS IN THE UNITED STATES IN 1860 IN SELECTED AGE GROUPS

District*	Age Groups			
	0-5	5-10	10-15	0-20
I	3,596	612	259	4,877
III	4,126	750	313	5,567
V	4,347	609	348	5,752
VII	4,767	592	384	6,219

*District I: Maine, New Hampshire, Vermont, Massachusetts, Rhode Island, Connecticut, New York.

District III: New Jersey, Pennsylvania.

District V: Delaware, Maryland, District of Columbia, Virginia, North Carolina.

District VII: South Carolina, Georgia, Florida, Alabama.

Source: U. S. Census of Population, 8th Census, <u>Mortality and Miscellaneous Statistics</u>, 1860.

Approximately 80-85% of the deaths occurring before age 20 are in the 0-10 year age group, and therefore the assumed average age of those dying in the 0-7.5 year group can be considered a lower bound estimate. However since the data portray the mortality conditions in 1860, the end of the period under study, the figures for the four districts on which the estimate is based will overstate the average age of those dying under 7.5 years of age for the earlier years. These two sources of bias should offset each other and give a fairly accurate portrayal of the average age of persons dying under 7.5 years of age for the entire period 1790-1860.

The actual subsistence costs in terms of social resources can be approximated by viewing some estimates of slave maintenance costs. Conrad and Meyer use a nursery cost per successful pregnancy of $50.[18] No doubt there is some imputation of the value of the mother's services as a field hand that were foregone for some portion of the pregnancy. Since the computations that follow do not purport to deal with slaves, and since the data on female wage rates and participation in the labor force are fragmentary; the opportunity costs of child-bearing are impossible to calculate. Such an opportunity cost does, however, conceptually belong in the nursery cost imputation. The assumption used in the calculation is that the social cost of subsistence for the first two years of life is $20 per annum.

The costs of maintenance from the second through the tenth year of life are approximated from several sources. Conrad and Meyer show that the cash costs of food and clothing in the period 1840-60 ranged between $25-$40, with an additional cost of $1.50-$2.00 per hand for medical expenses.[19] In 1855 DeBow's Review estimated yearly costs of maintenance and superintendence as, "$20 for clothing; $15 for corn, molasses and tobacco; $1 for physician's fees; $10 for overseer's wages and $15 for tools and repairs."[20] Kenneth Stampp stated that in the ante-bellum period "maintenance costs seldom exceeded $35."[21] It appears that by mid-period slave maintenance costs were in the $25-$40 range per annum. While these estimates are for full-grown slaves, two important cost elements are omitted, the cost of education to a modicum of literacy and the costs of shelter, that were no doubt a significant element of social cost in raising free population to productive age.[22] Since slave budgets represent a depressing lower bound estimate of subsistence costs, i.e., the economic quality of the free American population would have been lower if they had been raised to productive age on slave maintenance budgets, they should be raised by a small amount to represent a slight, apparent elevation of the subsistence datum among the free population. A social maintenance cost at or near the subsistence level for freemen from ages 3 through 10 of $40 per annum is used in the subsequent calculations. While the $40 maintenance cost estimate may seem large in comparison with slave maintenance budgets, this figure should also be seen in relation to the estimates of per capita income for this period. Parker and Whartenby have shown Martin's estimates of declining per capita income in the period 1799-1839 are in error on several counts.[23] Yet the Martin figures remain our only national income series for the period 1799-1838. Parker and Whartenby also found Kuznet's evidence "too weak" to support his claim of rising per capita income in the 1800 to 1840 period.[24] Paul David's estimates of per capita gross domestic product are the only unscathed numbers before 1840.[25] David's estimates show a rising trend of real per capita gross domestic product from 1800 to 1840. Although no dollar estimates of per capita income are derived in this work there is an implicit estimate in Tables 3.2 and 3.5 for 1840 of $102 per capita GNP (comprehensive) and $90 per capita GNP (restrictive) for 1839-40 in current dollars. The much criticized Martin data show three estimates of realized per capita income in current dollars for 1839: $98, $97, $95. David shows an increase in the index of real per capita gross domestic product of approximately 1-1/4% per annum from 1800 to 1840.[26] This would imply that per capita GNP was about $50 in 1790 in the prices of 1840.[27]

Thus it is seen that the estimate of $40 for maintenance costs over the period from 1790-1860 does not exceed any known estimate of per capita real GNP, and is considerably below any available estimates on per capita income in the 1800-1840 period, e.g., the assumed maintenance cost is only 45% of David's implicit per capita GNP in current prices in 1840. The maintenance cost estimate is then bounded from below by our knowledge of slave budgets and from above (though substantially below this upper limit) by available estimates of per capita income. The narrow differential between these maintenance costs and the estimated per capita GNP in 1790, $10 per head, is in rough accord with notions about the low level of savings in emergent economies.

If the annual social saving is to be computed in current dollars of each year some additional refinements would be required in the present calculation to account for changes in the price level. For example the social savings in 1850 would be the price-adjusted cost of maintenance (1840 = 100) in each of the preceding years valued in 1850 prices, etc.[28]

The computations in Table 3.3 use a maintenance cost estimate of $40 per annum throughout the period 1790-1860. If the social savings estimates are to be interpreted as occurring in the years immigration took place at current prices, how much error is introduced into the estimates, and what is the direction of these biases?

The appropriate deflator for the estimate would be a cost-of-living index; however the Federal Reserve Bank of New York index only begins in 1820. On the other hand the Warren-Pearson all commodity index extends backward to the middle of the eighteenth century.[29] Unless there were significant changes in distribution over the period 1820-1860 one would expect to find a high degree of association between the two indices. This is in fact the case and thus we infer changes in the cost-of-living index in the 1790-1820 period by observing the movement in the Warren-Pearson series on wholesale prices in order to form an estimate of the error introduced by assuming a constant dollar magnitude for maintenance cost when we want to interpret the social savings estimates as current dollar magnitudes in the year of immigration for each year in the period.[30]

From 1795-1819 the wholesale price index is an average of 30% above its 1840 level. The use of maintenance cost in 1840 prices understates the current dollar magnitudes of social savings calculations based on this period by the same percentage. From 1820 to 1835 there is an upward bias in the current dollar estimates based on this period of about 6%. From 1836-1840 there is a downward bias of 15%. From 1841-1850 the upward bias is 10% and for the remainder of the period there is a downward bias in the current dollar estimates of 6%. Over the entire period the average annual percentage of absolute deviation from the 1840 price level is 8%, with 16 years inducing downward bias, 20 years inducing upward bias, and 5 years of no bias. The high price level in the years spanning the hostilities with France and England are a source of large error leading to a downward bias of about 30% in the current dollar estimates in the period preceding 1820.

However, since the immigration flows were very small in this period, compared to the later years, our conclusion about the magnitude of resources potentially free for alternative uses due to the social savings attributable to immigration over the entire period 1790-1860 will not be altered materially by this source of error. In view of the above discussion it should be born in mind that the social savings estimates as current dollar magnitudes are subject to these errors and are thus crude approximations to the true current dollar magnitudes having variations of from $\pm 10\%$ for the periods indicated above.

The last assumption concerns the productive age itself. The 1860 Census refers to the 15-40 age group as the "responsible and productive period of life." However, scattered accounts of conditions in the early nineteenth century reveal that children were employed at various tasks well before their fifteenth year. The productive age for the period is probably on the interval between 10 and 15 years. The calculations take this lower limit of 10 years as the productive age. The uniform age distribution in the 0-15 age group makes the social saving calculation for this category invariant with respect to any choice of the productive age over 7.5 years. However, over the period

1790-1860 approximately 70%-80% of the immigrants were in the over-15 category. Thus the use of 10 years, instead of say 15 as the assumed age at which children are productive will yield an understatement of from 35%-40% in the social savings calculations. Clearly the choice of the productive age is one of the most crucial assumptions in the calculation.[31]

The magnitude of the social savings can be seen more clearly in the context of their potential for economic growth if we compare them with Gallman's estimate of gross capital formation shown in Table 3.2.

TABLE 3.2. GROSS CAPITAL FORMATION, SINGLE-YEAR ESTIMATES AND ESTIMATED SOCIAL SAVINGS CURRENT PRICES

(billion dollars)

Year	Manufacturing Durables	Total Annual GCF	Estimated Social Savings
	1	2	3
1839	.032	.200	.031
1844	.048	.187	.029
1849	.076	.257	.108
1854	.131	.541	.157
1859	.140	.532	.045

Cols. 2,3: Robert E. Gallman, "Gross National Product in the U. S., 1834-1909," in _Output, Employment and Productivity in the U. S. after 1800_, Studies in Income and Wealth, Vol. 30, NBER, 1966. Table A-3, page 34.

While the single year estimates of the social saving due to immigration never exceeded 40% of total gross capital formation in the U. S. (1849) it is noteworthy that for two years in the period 1839-1859 social saving exceeds gross capital formation in manufacturing durables (1849 and 1854) and is equal to the latter magnitude in 1839. Immigrant remittances over the period should be netted out of the social savings estimates and currency importations by immigrants should be added to give a more precise picture of the potential resource base for capital formation due to immigration. Data on assets brought in by immigrants have not been found for this period. Remittances averaged $400,000 per year from 1831-1840, $1,800,000 from 1841-1850, and $7,800,000 from 1851-1860.[32] Thus remittances would not serve to scale down the social savings plus net asset flows appreciably, in fact over the period 1851-1860 these remittances would be offset by net asset importations of less than 30 cents per immigrant.

As has been alluded to up to this point the significance of the pool of resources created out of the social savings attributable to immigration is to be seen in terms of the implication for capital formation.

The creation of the pool can be understood by reference to the conditions of population growth in two economies. In country A there is no immigration after, say 1790, thus the net increase in population is entirely made up of native reproduction. In country B, assume there is some annual net inflow of population and that in each case the level and rate of increase of population is equal to that in A. Then B will devote fewer resources to the maintenance of population in its unproductive years. The pool of resources that is committed to child raising in country A is greater than that of B. This differential resource requirement arising in two countries whose populations grow at the same rate is the social savings arising from immigration to country B. The question now arises as to the implication of this pool of resources for capital formation. If country B spends the entire amount of social savings on consumption, then there is no necessary difference in capital formation in the two countries, i.e., MPC = 1. If B re-invests the entire amount of free resources, MPS = 1, then B's rate of capital formation will grow by the compounded value of the annual social savings, i.e., all social savings are re-invested. These two cases, MPC = 1 and MPC = 0, are the extremes in terms of the implications for capital formation. On the other hand we may assume that the social earnings will raise disposable income (after necessaries) in the hands of households; and that an increment in uncommitted income will be spent in part for consumption and in part for savings.

Column 11 of Table 3.3 shows the compounded value at 6% per annum, in 1860 prices, of that increment of social savings going to capital formation. Gallman's estimates of capital formation as a percentage of GNP have been used as an approximate measure of the average propensity to save for the economy as a whole. If the APS, approximately 15%, also obtained out of social savings, then this percentage of free resources were directed annually toward investment. While it may be argued that marginal propensities to save are more appropriate, these would be above the APS, and imply something extra-marginal about social savings from immigration that need not be. Clearly the use of a MPS would increase the annual percentage of resources invested. Using the APS, and recognizing that it may be a source of downward bias, we can then calculate the economic saving (income minus consumption) out of the resource pool created by the social saving due to immigration. By 1860 the compounded value of the investment amounts to 846.5 million in 1860 prices or 94.5% of the estimated manufacturing capital stock, or 8.4% of the total capital stock.[33]

Thus the social savings due to immigration speeded up enormously the process of capital formation and growth in the American economy. At a 6% rate of increase in capital formation this finding implies that the American economy would have been retarded by almost one and one-half years, i.e., the 1860 capital stock and output would not have been reached until mid-1862 in the absence of immigration. Viewed in the light of this evidence the American growth experience was dependent on immigration for more than just skills and labor supply. The international flow of human beings to America was responsible for the formation of a large portion of the physical capital base of the economy and quickened the pace of development.

Human Capital Inflows to the American Economy: 1790-1860

While the relatively recent work of Conrad and Meyer has caused no small reaction to the concept of putting monetary estimates on human life, this valuation itself has a long history ante-dating the reign

TABLE 3.3. ESTIMATED NET IMMIGRATION, AGE DISTRIBUTION AND SOCIAL SAVINGS: 1790-1860

	1	2	3	4	5 = 1 x 2	6 = 1 x 3	7 = 1 x 4	8	9	10	11
Year	Est. Net Immigration (000)	% Under 15 Years	% 15-20 Years	% Over 40 Years	Number Under 15 Years (000)	Number 15-40 Years (000)	Number Over 40 Years (000)	Under 15 Man-Year-$ Equivalent (000,000)	Over 15 Man-Year-$ Equivalent (000,000)	Total Social Savings (000,000)	Implied GCF (1860=100) (000,000)
1790	3.5				1.0	2.2	.3	.284	.973	1.257	12.1
91	4.0				1.1	2.5	.4	.325	1.112	1.435	14.7
92	5.0				1.4	3.1	.5	.406	1.390	1.795	14.7
93	3.6				1.0	2.2	.4	.292	1.001	1.292	10.1
94	3.5				1.0	2.2	.3	.284	.973	1.257	8.7
95	3.9				1.1	2.4	.4	.317	1.084	1.401	7.9
96	4.5				1.3	2.8	.4	.366	1.251	1.615	7.9
97	3.5				1.0	2.2	.3	.284	.973	1.292	6.5
98	3.8				1.1	2.3	.4	.309	1.056	1.365	6.8
99	4.0				1.1	2.5	.4	.325	1.112	1.435	6.8
1800	3.8	28.4			1.1	2.3	.4	.309	1.056	1.265	5.7
01	4.0		61.6		1.1	2.5	.4	.325	1.112	1.435	5.6
02	4.5				1.3	2.8	.4	.366	1.251	1.615	6.6
03	3.9			10.0	1.1	2.4	.4	.317	1.084	1.401	5.4
04	9.5				2.7	5.9	1.0	.772	2.641	3.412	11.6
05											1.6
06											1.6
07											1.5
08											1.6
09	1.5				.4	.9	.2	.122	.417	.539	1.4
1810											1.3
11											1.2
12											1.1
13											.9
14											.7
15											.8
16	9.0				2.6	5.6	.9	.733	2.507	3.240	4.8
17	16.7				4.7	10.3	1.7	1.355	4.637	5.992	8.4
18	12.4				3.5	7.7	1.2	1.010	3.457	4.465	6.0
19	10.7				3.0	6.6	1.1	.867	2.967	3.834	5.5
1820	8.4	26.6	58.8	14.6	2.2	4.9	1.2	.634	2.389	3.023	4.6
21	9.1	27.5	60.5	12.0	2.5	5.5	1.1	.718	2.569	3.286	5.0
22	6.9	25.3	63.5	11.2	1.7	4.4	.8	.500	2.004	2.504	3.4
23	6.4	23.8	64.3	11.9	1.5	4.1	.8	.432	1.880	2.311	3.2
24	7.9	20.5	68.0	11.5	1.6	5.4	.9	.464	2.442	2.906	4.1

TABLE 3.3 (Continued)

	1	2	3	4	5 = 1 x 2	6 = 1 x 3	7 = 1 x 4	8	9	10	11
Year	Est. Net Immigration (000)	% Under 15 Years	% 15-20 Years	% Over 40 Years	Number Under 15 Years (000)	Number 15-40 Years (000)	Number Over 40 Years (000)	Under 15 Man-Year-$ Equivalent (000,000)	Over 15 Man-Year-$ Equivalent (000,000)	Total Social Savings (000,000)	Implied GCF (1860=100) (000,000)
1825	10.2	18.0	73.0	9.0	1.8	7.4	.9	.525	3.247	3.771	4.8
26	10.8	18.7	72.1	9.2	2.0	7.8	1.0	.579	3.420	3.999	5.1
27	18.9	25.5	64.6	9.9	4.8	12.2	1.9	1.377	5.459	6.836	7.9
28	27.4	29.0	60.9	10.1	7.9	16.7	2.8	2.271	7.548	9.819	10.8
29	22.5	45.5	47.3	7.2	10.2	10.7	1.6	2.931	4.765	7.695	7.9
1830	23.3	69.8	25.5	4.7	16.3	5.9	1.1	4.656	2.734	7.390	7.7
31	22.6	35.3	56.9	7.8	8.0	12.9	1.8	2.285	5.685	7.970	7.5
32	60.5	34.9	57.2	7.9	21.1	34.6	4.8	6.037	15.287	21.324	18.5
33	58.6	33.5	58.4	8.1	19.6	34.2	4.8	5.618	15.143	20.761	16.7
34	65.4	27.0	63.0	10.0	17.6	41.2	6.5	5.048	18.526	23.574	17.6
35	45.3	22.4	66.5	11.1	10.2	30.2	5.0	2.907	13.670	16.577	11.7
36	76.2	22.3	67.6	10.1	17.0	51.5	7.7	4.863	22.999	27.862	17.2
37	79.3	26.2	63.9	9.9	20.8	50.7	7.8	5.945	22.733	28.678	15.8
38	38.9	23.7	63.6	12.7	9.2	24.8	4.9	2.638	11.527	14.165	7.3
39	68.1	22.0	68.4	9.6	15.0	46.6	6.5	4.283	20.613	24.896	12.0
1840	84.1	24.1	67.7	8.2	20.3	56.9	6.9	5.794	24.772	30.566	17.0
41	80.3	23.2	67.0	9.8	18.6	53.8	7.9	5.327	23.940	29.267	15.3
42	104.6	24.2	67.1	8.7	25.3	70.2	9.1	7.237	30.772	38.009	20.4
43	52.5	29.6	61.2	9.2	15.5	32.1	4.8	4.444	14.349	18.793	10.5
44	78.6	25.2	64.6	10.2	19.8	50.8	8.0	5.666	22.830	28.496	14.4
45	114.4	23.6	66.3	10.1	27.0	75.8	11.6	7.720	33.924	41.644	19.0
46	154.4	24.1	65.1	10.8	37.2	100.5	16.7	10.643	45.503	56.146	21.2
47	235.0	25.9	65.4	8.7	60.9	153.7	20.4	17.405	67.597	85.002	30.3
48	226.5	24.0	65.9	10.1	54.4	149.3	22.9	15.549	66.839	82.388	29.7
49	297.0	22.8	67.0	10.2	67.7	199.0	30.3	19.368	89.024	108.372	41.4
1850	370.0	21.8	66.6	11.6	80.7	246.4	42.9	23.068	112.328	135.396	46.3
51	379.5	22.1	67.1	10.8	83.9	254.6	41.0	23.985	114.765	138.750	40.5
52	371.6	27.2	61.9	10.9	101.1	230.0	40.5	28.908	105.029	133.937	36.7
53	368.6	22.1	66.8	11.1	81.5	246.3	40.9	23.301	111.493	134.794	33.1
54	427.8	21.9	67.8	10.3	93.7	290.1	44.0	26.797	129.726	156.521	38.0
55	200.9	23.4	65.7	10.9	47.0	132.0	21.9	13.443	59.739	73.182	16.0
56	200.4	27.9	63.2	8.9	55.9	126.7	17.8	15.994	56.107	72.101	14.7
57	251.3	26.5	65.1	8.4	66.6	163.6	21.1	19.046	71.712	90.758	16.9
58	123.1	18.3	71.0	10.7	22.5	87.4	13.2	6.444	39.054	45.498	8.1
59	121.3	16.2	73.4	10.4	19.6	89.0	12.6	5.619	39.458	45.077	7.9
1860	153.6	16.2	74.5	9.3	24.9	114.5	14.2	7.119	49.986	57.105	9.7
											846.5

Notes to Table 3.3:

Column 1:

Net immigration for 1817 is taken as 75% of the immigration reported in Bromwell, op. cit., of 22,240. This latter figure includes returning American citizens, the 75% adjustment is less than the ratio of net to gross immigration for the decade 1820-30.

The estimates for the period 1805-15 are based on Bromwell's discussion of the impact of the European hostilities on immigration to America. In 1815 the U. S. Census of Population, 8th Census, 1860, p. xxv reports immigration from "Great Britain and Ireland" alone of 1,209. While peace was concluded between Britain and the U. S. in February of 1815, it was almost a year before immigration continued on an extensive scale. Thus the figure for 1815, slightly increased to allow for immigration from other countries, is used as typical of the 1805-1815 period.

Immigration for 1818 and 1819 is based on the 8th Census, op. cit.

For the period 1820-1860 the figures on immigration exclusive of returning American citizens are taken from Historical Statistics, op. cit., page 57, Table C 98-114. The ratio of immigration exclusive of returning American citizens to immigration including returning citizens as reported in the Historical Statistics exceeds Willcox's estimated adjustment factors. Walter F. Willcox, Studies in American Demography, Cornell University Press, 1940, page 390 reports the following percentages.

Period	Assumed percent that net immigration was of gross
1820-30	100
1831-40	95
1841-50	90
1851-60	85

If the census reported foreign born population in the early years of the period 1790-1860 net immigration could be solved for analytically by using our knowledge of the age distribution of gross immigration as an approximation to the age distribution of net immigration and death rates applicable to the various age classes by the following formula:

$$(FBP_1) = (FBP_0) + \sum_{j,k=1}^{m\ 10} NI_{jk} - \left[\sum_{i}^{n} \sum_{j}^{m} (FBP_0)_j \cdot d_i + \sum_{i}^{n} \sum_{j}^{m} \sum_{k=1}^{10} NI_{j,k} \cdot d_i \right]$$

where,

FBP_0 = foreign born population reported in the census at the beginning of the decade

FBP_1 = foreign born population reported in the census at the end of the decade

d = age specific death rate

j = index over the age distributions

k = index over the years in the decade 1, . . ., 10

i = index over the death rates per 10,000 applicable over the age distributions

NI = net immigration

thus,

*net, live immigration over the decade $= \sum_{j}^{m} \sum_{k=1}^{10} NI_{jk}$

$$\frac{FBP_1 - FBP_0 + \sum_{i}^{n} \sum_{j}^{m} (FBP_0)_j \cdot d_i}{1 - \sum_{i}^{n} d_i}$$

Columns 2, 3 and 4:

These columns are derived from the <u>Historical Statistics</u>, <u>op</u>. <u>cit</u>., page 62, Table C 133-138. The "age not reported" category has been included in the under 15 years category.

Column 5:

Col. 5 = Col. 1 X Col. 2.

Column 6:

Col. 6 = Col. 1 X Col. 3.

Column 7:

Col. 7 = Col. 1 X Col. 4.

Column 8:

Uniform age distribution over the 0-15 age bracket gives the average age of an immigrant in this group as 7.5 years. From U. S. Census, <u>Mortality and Miscellaneous Statistics</u>, <u>op</u>. <u>cit</u>., the average death rate for Districts I, III, V and VII in the 0-7.5 age group is 4,634 per 10,000 entering this group. The average age at the time of death is 18 months. For every 10,000 entering the 0-7.5 group only 5,366 will survive to age 7.5 and beyond, those who die must be supported for 18 months of life. Assuming a cost to the economy of $20 per annum for the first two years of life, and $40 per annum thereafter, the embodied social cost of a person 7.5 years of age is given by the following formula:

$$ESC_{7.5} = \frac{AC_{0-2}(AATD_{0-7.5})(D_{0-7.5})}{S_{7.5}} + \frac{\left[2AC_{0-2} + 5.5AC_{3-7.5}\ S_{7.5}\right]}{S_{7.5}}$$

where,

$ESC_{7.5}$	=	embodied social cost in a 7.5 year old
AC_{0-2}	=	avg. annual maintenance cost for first two years of life
$AC_{3-7.5}$	=	avg. annual maintenance cost for years 3 through 7.5
$AATD_{0-7.5}$	=	avg. age at time of death for those who die on the 0-7.5 interval
$D_{0-7.5}$	=	number of deaths on the 0-7.5 interval
$S_{7.5}$	=	number who survive to age 7.5 and beyond

or,

$$ESC_{7.5} = \$285.90$$

$$= \frac{\$20(1.5)(4,634) + \$260(5,366)}{5,366}$$

and,

$$\text{Col. } 8 = ESC_{7.5} \text{ X Col. } 5.$$

Column 9:

These groups are comprised of individuals who have survived the 10th year of life, our assumed lower limit for entry in the productive period. For each 10,000 who enter the 0-10 age group 4,849 will die before reaching age 10. Employing the same assumption as to average age at time of death on this interval as in Col. 8, the social cost (saving) embodied in a person in this group is:

$$ESC_{10.0} = \$388.24$$

$$= \frac{\$20(1.5)(4,849) + \$360(5,151)}{5,151}$$

and,

$$\text{Col. } 9 = (\text{Col. } 6 + \text{Col. } 7) \text{ X } ESC_{10.0}$$

Column 10:

Col. 10 = Col. 8 + Col. 9.

Column 11:

R. E. Gallman, "Gross National Product 1834-1909," op. cit., p. 11, Table 3. Gallman's shares of capital formation in GNP*II are used from 1834-1860 as an estimate of the investment undertaken out of the resource pool created by the social savings. Before 1834 a 15% share of capital formation in GNP is used. The annual investments made out of the social savings fund are then converted to the prices of 1860 and compounded to this date at a 6% rate. The social savings figures are deflated by the Federal Reserve Bank, Cost-of-Living Index, Historical Statistics, op. cit., Series E 157-160, page 127, from 1820-1860. From 1790-1819 the index is estimated by the regression described in footnote (30).

of Alfred the Great.[34] Under old Anglo-Saxon law a wer or wergild was set on human life based on social position. If a person was killed in battle or by a homicide the killer was required to pay a wer to the surviving relatives. The wer itself was an arbitrarily chosen estimate of the capital value of a man. The first explicit formulation of the value of a human being was probably set down by Sir William Petty (1623-1687) in his "Political Arithmetick." Petty's method subtracted from total national income the earnings of land and capital, the residual, total returns to labor, was divided by the population to form an estimate of "the value of each Head of Man, Woman and Child." The basis of the modern approach to human capital, i.e., the capitalization of the earnings stream is suggested in Adam Smith. The fourth type of fixed capital is the "acquired and useful abilities" of the members of a society.

> Fourthly, of the acquired and useful abilities of all the inhabitants or members of society. The acquisition of such talents, by the maintenance of the acquirer during his education, study, or apprenticeship, always costs a real expense, which is a capital fixed and realized, as it were, in his person. Those talents, as they make a part of his fortune, so do they likewise of that of the society to which he belongs. The improved dexterity of the workman may be considered in the same light as a machine or instrument of

> trade which facilitates and abridges labor, and which, though it costs a certain expense, repays that expense with a profit.[35]

Throughout the nineteenth and early twentieth centuries other prominent classical economists, J. S. Mill, Alfred Marshall and Irving Fisher, were elaborating the concept of Adam Smith's fourth type of fixed capital.[36] For the purposes of this section it is sufficient to point out that this tradition has become firmly embedded in the modern literature with the work of Gary Becker.[37]

One previous work has attempted to merge the concept of human capital with immigration flows.[38] Based on the work of Dr. Agostino DeVita, Gini has estimated the compounded value of human capital due to immigration at six trillion dollars in 1930 prices. Besides the use of the dubious theoretical concept of continuous compounding of the single year estimates of human capital inflows, there are other points in execution which render the estimate implausible.[39]

The first is the obvious blunder of assuming that the participation rate of all immigrants (men and women) in the labor force is 100%. Secondly, the use of average annual income of $1,500 in 1930 includes returns to real property as well as wage income. This figure implies average wage incomes, the relevant concept for human capital calculations, of around $800 in the 1820-1860 period in the prices of 1840 or more than three times greater than the estimate used in the following calculation.

The procedure for this estimate of single year inflows of gross human capital will be made explicit with a full discussion of the assumptions employed in the model and their documentary and logical basis in addition to an analysis of sources of error.

The assumptions include the following:

a) The occupational distribution of the immigrant labor force.

b) The ratio of labor force to population applicable to the annual net immigrations.

c) The wage differentials among the respective occupational groups and the earnings rate of common, free labor.

d) The rate of discount applicable to future earnings.

e) The age distribution of immigrants and the productive life span.

Assumption a) is most difficult to support in the period 1790-1819, since after 1820 we have reasonably accurate estimates of the occupational distribution of immigrants.[40] Of the seven reported occupational categories, professional and commercial have been grouped together; skilled labor is a separate category; and farmers, laborers and servants comprise the third category. To estimate the three-way distribution of occupations in the period 1790-1819 it has been assumed that the ratio of each category to net immigration was the same in this period as in the period 1820-1830.[41]

It may be seen in Table 3.4 that the ratio of productive immigrants, i.e., those declaring occupations, to total net immigrants is not significantly above the ratio of total labor force to American population.[42] Given, however, that 70%-80% of the immigrants were in the over-15 year age bracket we might expect a higher potential participation rate among the immigrants. Although the immigration figures on servants include women, the category is small in relation to the total of all productive groups, and it is most probably the case that the data from Table C 115-132, Historical Statistics, understate the participation of immigrant women since none of the remaining occupational categories would be likely to include women

TABLE 3.4. ESTIMATED PARTICIPATION RATES FOR NATIVE AND FOREIGN BORN POPULATION, 1800-1860

1	2	3
Year	Labor Force to Total Population (%)	Productive Groups to Net Immigration (%)
1800	35.9	39.5
1810	32.4	39.2
1820	32.5	33.6
1830	32.6	18.4
1840	33.0	48.6
1850	35.5	33.3
1860	35.3	47.6

Notes:

Col. 2 : Stanley Lebergott, "Labor Force and Employment, 1800-1960," in Output, Employment and Productivity . . ., op. cit., p. 118, Table 1.

Col. 3 : Based on Tables 1 and 2.

(professional, commercial, skilled, farmers and laborers). Thus the participation rate of immigrants is downward biased by the omission of women in categories other than "servant." Another source of downward bias is the omission of the "no occupation reported" group from the human capital estimate since this category did contain some unknown percentage of individuals whose occupations were not reported.[43]

Table 3.5 shows the decade average, daily wage paid to non-agricultural labor.

TABLE 3.5. DECADAL AVERAGE DAILY WAGE OF NON-AGRICULTURAL LABORERS 1790-1880

1	2	3	4 = 2 x 3
Decade	Average Daily Wage ($)	Working Days/Year	Annual Wage Income of Common Labor ($)
1790-1799	.564	250	$141
1800-1809	.819	250	205
1810-1819	.949	250	237
1820-1829	.780	250	195
1830-1839	.979	250	245
1840-1849	1.00	250	250
1850-1859	1.00	250	250
1860-1869	1.40	250	350
1870-1880	1.20	250	300

Notes:

Column 2 is based on the "medium" estimates for Massachusetts contained in Carroll D. Wright, Comparative Wages, Prices and Cost of Living, Report of the Massachusetts Bureau of the Statistics of Labor, 1885, pp. 53-54. Where only "high" and "low" figures are given the average has been used. Edward Young, Labor in Europe and America, Philadelphia, 1875, page 744 shows the wages of common labor advancing by 54% between 1860 and 1870. C. D. Wright, op. cit., pp. 226-227, indicates that the wages of agricultural labor advanced 25% from 1860-1872. In the above table an increase of 40% in the wages of common labor from 1860-1869 is used. S. Lebergott, "Wage Trends, 1800-1900," op. cit., p. 467, indicates that wages in the North

Notes to Table 3.5 - Cont.:

rose 55%-66% during the Civil War. Lebergott, op. cit., p. 462, also gives daily wage trends in the U. S. for selected years as follows:

Year	Non-farm Labor
1800	$1.00
1812-15	1.00
1818	0.75
1830-32	0.75
1840	0.85
1850	0.90
1860	1.04
1870	1.57
1880	1.28

From 1870-1880, C. D. Weight, op. cit., p. 239, shows a decline of the average weekly wage paid to all employees in Massachusetts of 15%. Lebergott, op. cit., p. 407, shows a decline of 12.5% for the wages of non-farm labor over the same period. The data in Table 3.5 show a 15% decline in this period.

While the Lebergott wage data for non-farm labor in selected years given above appear at variance with the decadal average in Table 3.5 over the period 1830-1850, the discrepancy arises because of comparison of decade averages with single year estimates and Lebergott's single year estimates correspond to the data for those years included in the decade averages of Table 3.5.

Column 3 :

An average of 250 working days per year is assumed.

Column 4 :

4 = 3 x 2 . Conrad and Meyer, op. cit., give the annual hiring rates for male slaves in 1860:

State	Men
Virginia	$105
North Carolina	110
South Carolina	103
Georgia	124
Florida	139
Alabama	138
Mississippi	166
Louisiana	171
Texas	166
Arkansas	170
Tennessee	121

While the comparisons between free and slave labor are tenuous at best, it may be noted that slave rental rates in the new South are about two-thirds the calculated earnings rates for free labor in 1860. One would need to know more about the extent of imperfections in the markets for Negro labor services to use Conrad and Meyer's estimates as a benchmark.

Given that the data in Table 3.5 are based on wage returns in Massachusetts, there probably is an upward bias here, although curiously enough they are comparable to Lebergott's figures for the U. S. as a whole. The decadal averages are based on single year estimates given in the current prices of that year. Ideally, the discounting of wage incomes should be preceded by deflating the wage estimates over the productive span by the wage index for the year of entry of the immigrant. Thus current wages in 1850 would be deflated by the wage index in 1840 for an immigrant laborer arriving in 1840. The 1850 current wage earnings would be deflated by the 1841 wage index for an immigrant arriving in 1841, etc. This procedure would give human capital estimates for the year the immigrant arrived based on current prices in that year. Due to the lack of a reliable index before 1860, and the tentative nature of the earnings estimates in Table 3.5, this last procedure, while conceptually desirable, is not empirically possible. Given the state of our knowledge about real wages before 1860, such a refinement would lend bogus accuracy to the wage earnings estimates, which in actuality are of a rough and ready sort.[44]

The earnings differential between skilled and unskilled labor is taken as 200%.[45] Based on data in Zachariah Allen's Science of Mechanics, Rosenberg has estimated this differential for the 1820's, and it appears to be roughly correct for the period 1790-1860. This differential is also obtained over the period when sample calculations are made on the data given in C. D. Wright's work on wage rates in Massachusetts. The differential between the wages of common labor and the professional-commercial category is assumed to be 250%.[46]

The discount rate at which earnings are capitalized will affect the human capital estimates crucially. Following Conrad and Meyer, a 6% rate of discount is used in the estimates. If an 8% rate were used it would lower the single year estimates of human capital by approximately 14%, a 4% rate would raise them 17%-18%.

The modal age of immigrants was 20-30 between 1820-1855.[47] Table 3.6 shows the years of additional life for individuals of various ages in 1818 and 1830. As is discussed in the "notes" to this table the use of a productive age span based on the Pennsylvania Life Tables estimates for 1818 tends to give a

TABLE 3.6. YEARS OF ADDITIONAL LIFE AT VARIOUS AGES, 1818 AND 1850

1	2	3
Age	Years of Additional Life, 1818	Years of Additional Life, 1850
20	27.04	41.46
25	24.14	37.86
27	23.08	- -
28	22.55	- -
30	21.48	34.34
35	19.95	- -
40	19.15	- -
45	17.91	- -

Notes:

Column 2 :

Adam Seybert, *op. cit.*, p. 51. Seybert's data are from the "Pennsylvania Company for Insurance in Lives and Granting Annuities" (board of health estimates).

Column 3 :

J.D.B. DeBow, *Compendium of the Seventh Census*, Washington, 1854, p. 122. These estimates are based on the Carlisle Life Table. Dublin and Lotka, *Length of Life*, The Ronald Press, 1936, p. 40, points out that the estimates for 1818 are of "questionable reliability" and the expectation of life from the Pennsylvania Table at age 1 is given as 25.96 years. In summarizing various early American life tables (Meech, Farr, Billings, Elliot, etc.). Dublin and Lotka sug- gest that the expectation of life at birth in the U. S. at the beginning of the nineteenth century was between 35-40 years. Thus if the Pennsylvania Insurance Table understates the additional years of life by the amount it is below the Dublin-Lotka average expectation at birth, the use of this life table is a source of at least a 40% downward bias in the extimates of produc- tive life span, and would tend to underestimate the value of human capital by 17%-18% in each year.

17%-18% downward bias to the human capital estimates. Thus the assumption of a 20 year productive span for each incoming migrant at an average age of 25 years is an extremely conservative one. Although the expected years of additional life at age 25 is 24.14 years in Table 3.6, it is assumed that the last 4 years of expected life are not productive.

Column 9 of Table 3.8 shows the conjectural estimates of the value of human capital embodied in the annual immigration to the United States from 1790-1860 derived on the basis of the evidence presented in this section. Table 3.7 shows the percentage that these estimates were of total gross capital formation for selected years from 1839-1859. Column 3 of Table 3.7 shows the impact of adjusting the human capital estimates so that they correspond to values in the prices of the year for which they are given. This adjustment is discussed in the notes to Table 3.7. After 1841 the advance of money prices gives a large upward bias to the estimates in terms of current price valuations. Even after the money wage adjustments, the human capital estimates average 80% of the single year estimates of total gross capital formation over the period 1839-1859. Bearing in mind the rather sharply downward biased assumption of a 20 year productive span, and the

TABLE 3.7. HUMAN CAPITAL AS A PERCENTAGE OF GROSS CAPITAL FORMATION (GCF), SINGLE-YEAR ESTIMATES FOR SELECTED YEARS, 1839-1859

1	2	3
Year	(Total human capital ÷ total GCF) X 100	(Adj. total human capital ÷ total GCF) X 100
1839	74%	74%
1844	77	54
1849	224	135
1854	170	85
1859	68	48

Notes:

Column 2 :

Robert E. Gallman, "Gross National Product in the U. S., 1834-1909," *op. cit.*, Table A-3, p. 34. Gross capital formation (GCF) single year estimates in current prices are based on Gallman's work.

Column 3 :

Recognizing Lebergott's caveat on wage indexes in the first half of the nineteenth century, it is nevertheless necessary to look at his estimates on the change in money wages over this period. The following data is taken from S. Lebergott, *op. cit.*, p. 493, Table 6.

Notes to Table 3.7 - Cont.:

Trends in Wages, U. S., Selected Periods 1800-1880 (% change)

	Money wages	
	Non-farm	Farm
1800-1818	-10 to 20	-5 to 10
1818-1830	(a)	-5 to 10
1830-1840	+10 to 20	+15 to 20
1840-1850	(a)	(a)
1850-1860	+5 to 15	+20 to 30
1860-1870	+60 to 70	(a)
1870-1880	-15 to 25	-15 to 25

(a) Change of less than 5%

Since the human capital estimates are intended to portray values in the current prices of the year of immigration a large positive or negative change in money wages on the productive life span (20 year horizon) will seriously distort the estimates. Since the productive span extends 20 years forward from the date of immigration, those immigrants entering in 1841 will have their earnings in the last year of their span valued in money wages in 1861 prices.

As the above table shows money wages in the non-farm sector advanced from 65% -85% above 1840 levels over the 1860-1870 decade; in the farm sector the advance was roughly 20%-30%. Immigrants in 1850 will have the ten years from 1860-1870 of their productive spans valued in money prices 60%-70% above the 1850 price level, etc. Thus after 1841 there is a significant upward bias in the human capital estimates if these estimates are interpreted as valued in the current prices in the year of immigration. The following adjustment percentages have been calculated for 5-year intervals after 1841, by considering the timing of the money wage change on the productive interval and averaging Lebergott's data for the farm and non-farm sectors by the percentages of the labor force accounted for by these sectors. Before 1841 the indicated advances and declines in money wages will not affect the estimates by more than 10% and the adjustment is omitted.

Notes to Table 3.7 - Cont.:

1841-45	+30%
1846-50	+40%
1851-55	+50%
1856-60	+30%

low participation rate estimates for immigrants it is not too far off the mark to state that the single year estimates of human capital flows to the U. S. in current prices of the year of immigration are of the same order of magnitude as single year estimates of total gross physical capital formation over the period 1839-1859.

Having viewed the magnitude of the human capital flows, it might be well to assess their significance. The American economy has been thought of as capital intensive in terms of its production techniques. Intensity is a relative term and the significance of capital intensity for the development of American technology vis a vis British technique is to be found in relation to the appropriate measure of capital intensity between these two countries rather than in terms of some arbitrary standard of historical comparison, or worse, what might have been. It has been observed that technologies may be said to differ if the types of inputs are in some relevant sense different between two countries.[48]

In particular, Ames and Rosenberg, have raised serious questions about the appropriateness of using production functions with homogeneous inputs. In Chapter I the introduction of two types of labor into the production function produced results that would not be obtainable with the more conventional specification.

Table 3.8 shows that the human capital from the immigration of the combined professional, commercial and skilled categories exceeded that of common labor from 1790-1844. Table 3.9 displays the annual immigration of three occupational groups, mechanics, engineers, and manufacturers from 1820-1855. Bromwell does not state if these figures are to be construed as net immigration, but the totals: 341, 512 mechanics, 2,016 engineers, 3,120 manufacturers; are suggestive of a sizable transference of knowledge and skills to the American economy by the movement of people.[49]

While these data are interesting the question persists how can the American economy be characterized as capital intensive when the single year estimates of human capital from immigration alone are almost as large as those estimates of gross, physical capital formation? The notion that American production is capital intensive could easily arise, and persist, if the measure of intensity is that such as is employed in Col. 6, Table 2.5, Chapter 2, i.e. dollars of capital per worker. This measure representing the capitalized value of the marginal product of physical capital at

TABLE 3.8. CONJECTURAL ESTIMATE OF SINGLE YEAR FLOWS OF HUMAN CAPITAL INTO THE UNITED STATES BY IMMIGRATION: 1790-1860

	1	2	3	4	5	6	7	8	9
Year	Est. % Net Immi-gration to Gross	No. Pro-fessional and Commercial	No. Skilled	No. Farm-ers, Labor-ers and Servants	P.V. of Laborers Earnings a 6% ($)	$ Human Capital (Prof.-Com.) (000,000	$ Human Capital (Skilled) (000,000)	$ Human Capital (Farm, Lab. and Serv.) (000,000)	Total Gross Human Capital (000,000)
1790	↑	429	423	528	$1,881	2.017	1.591	.993	4.061
91	│	490	483	603	1,926	2.359	1.861	1.161	5.381
92	│	613	604	754	1,974	3.025	2.385	1.488	6.898
93	│	441	435	543	2,026	2.234	1.763	1.100	5.097
94	│	429	423	528	2,080	2.231	1.760	1.098	5.089
95	│	478	471	588	2,138	2.555	2.014	1.257	5.826
96	│	552	544	679	2,194	3.035	2.393	1.493	6.921
97	│	429	423	528	2,264	2.428	1.915	1.195	5.538
98	│	466	459	573	2,332	2.717	2.141	1.336	6.194
99	│	490	483	603	2,405	2.946	2.323	1.450	6.719
1800	│	466	459	573	2,483	2.893	2.279	1.423	6.595
01	│	490	483	603	2,490	3.050	2.405	1.501	6.956
02	│	552	544	679	2,492	3.439	2.711	1.692	7.842
03	80.0	478	478	588	2,498	2.985	2.353	1.469	6.807
04	│	1,165	1,148	1,433	2,504	7.292	5.749	3.588	16.629
05	│	↑	↑	↑	2,509	1.154	.908	.567	2.629
06	│	│	│	│	2,516	1.157	.911	.569	2.637
07	│	│	│	│	2,523	1.161	.913	.570	2.644
08	│	│	│	│	2,530	1.164	.916	.572	2.652
09	│	184	│	│	2,537	1.167	.918	.573	2.658
1810	│	│	181	226	2,545	1.170	.921	.575	2.666
11	│	│	│	│	2,542	1.169	.920	.574	2.663
12	│	│	│	│	2,529	1.163	.915	.572	2.650
13	│	│	│	│	2,521	1.160	.913	.570	2.643
14	│	│	│	│	2,511	1.155	.909	.567	2.631
15	│	↓	↓	↓	2,501	1.150	.905	.565	2.620
16	│	1,106	1,090	1,361	2,491	6.888	5.430	3.390	15.708
17	│	2,045	2,015	2,515	2,480	12.679	9.994	6.237	28.910
18	│	1,524	1,501	1,874	2,467	9.399	7.406	4.632	21.428
19	↓	1,309	1,289	1,610	2,456	8.037	6.332	3.954	18.323
1820	81.3	844	886	1,095	2,442	5.153	4.327	2.674	12.154
21	78.4	1,290	1,202	1,408	2,476	7.985	5.952	3.486	17.423
22	80.8	1,278	1,129	1,025	2,503	7.997	5.652	2.566	16.215
23	76.9	1,235	975	880	2,535	7.827	4.943	2.231	15.001
24	82.2	1,737	1,017	1,078	2,571	11.165	5.229	2.772	19.166

TABLE 3.8 (Continued)

	1	2	3	4	5	6	7	8	9
Year	Est. % Net Immi- gration to Gross	No. Pro- fessional and Commercial	No. Skilled	No. Farm- ers, Labor- ers and Servants	P.V. of Laborers Earnings a 6% ($)	$ Human Capital (Prof.-Com.) (000,000)	$ Human Capital (Skilled) (000,000)	$ Human Capital (Farm, Lab. and Serv.) (000,000)	Total Gross Human Capital (000,000)
1825	79.3	1,622	1,123	1,876	2,608	10.575	5.858	4.893	21.326
26	77.9	1,562	1,658	1,689	2,648	11.002	8.781	4.472	24.255
27	86.7	2,027	2,650	3,440	2,690	13.632	14.257	9.254	37.143
28	90.7	2,412	3,508	5,071	2,734	16.486	19.182	13.864	49.532
29	93.2	2,715	2,404	3,249	2,781	18.876	13.371	9.035	41.282
1830	93.9	1,468	1,639	2,034	2,830	10.386	9.277	5.756	25.419
31	94.8	2,418	2,259	3,534	2,833	17.125	12.799	10.012	39.936
32	98.1	5,494	10,137	11,655	2,836	38.952	57.497	33.054	129.503
33	97.9	4,810	12,531	10,582	2,840	34.151	71.176	30.053	135.380
34	96.2	3,446	6,917	10,842	2,843	24.492	39.330	30.824	94.646
35	93.1	4,061	5,591	8,950	2,846	28.896	31.824	25.472	86.192
36	94.2	3,628	8,364	16,540	2,849	25.813	47.658	47.122	120.593
37	93.4	4,124	7,923	18,727	2,854	29.425	45.224	53.447	128.096
38	86.2	3,848	4,892	8,959	2,858	27.494	27.963	25.605	81.062
39	91.2	5,724	9,144	18,586	2,863	40.970	52.359	53.212	146.541
1840	91.2	5,282	9,860	25,809	2,867	37.859	56.537	73.994	168.390
41	91.4	5,309	10,155	22,566	2,899	38.477	58.879	65.419	162.775
42	94.2	5,388	13,709	28,430	2,932	39.494	80.390	83.357	203.241
43	93.7	3,564	5,709	12,921	2,967	26.436	33.877	38.337	98.650
44	92.7	4,371	8,784	19,217	3,004	32.826	52.774	57.728	143.328
45	95.4	5,334	10,358	36,627	3,043	40.578	63.039	111.456	215.073
46	97.4	4,657	12,906	48,199	3,085	35.917	79.630	148.694	264.241
47	98.1	4,828	25,403	81,090	3,129	37.762	158.972	253.731	450.470
48	98.7	3,873	24,384	81,256	3,176	30.752	154.887	258.069	443.708
49	99.1	4,440	31,733	104,575	3,226	35.809	204.741	337.359	577.909
1850	97.2	7,114	25,631	90,120	3,278	58.299	168.037	295.413	521.749
51	92.8	14,775	33,684	152,038	3,320	122.633	223.662	507.754	854.049
52	93.5	11,289	25,410	125,507	3,362	94.884	170.857	421.955	687.696
53	91.9	12,410	19,121	131,676	3,407	105.702	130.290	448.620	684.612
54	92.9	14,745	33,879	160,641	3,454	127.323	234.036	554.854	916.213
55	87.2	13,550	15,227	69,648	3,505	118.338	106.741	244.116	469.195
56	89.3	10,326	16,786	56,696	3,558	91.850	119.449	201.724	413.023
57	92.3	11,707	24,055	73,087	3,617	105.861	174.014	264.356	544.231
58	85.0	9,247	15,931	37,370	3,677	85.003	117.157	137.409	339.569
59	78.0	10,415	19,210	30,654	3,741	97.406	143.729	114.677	355.812
1860	85.5	10,259	16,537	46,533	3,809	97.691	125.979	177.244	400.914

Notes to Table 3.8:

Column 1 :

Historical Statistics, <u>op</u>. <u>cit</u>., pp. 57 and 61.

Col. 1 = (Col. 88 series C 88-114 - Col. 115 series C 115-132) X 100

= percentage that immigration exclusive of returning citizens was to gross

Column 2 :

Col. 2 = (series C 115-132, Cols. 126 + 127) X Col. 1.

Column 3 :

Col. 3 = (series C 115-132, Col. 128) X Col. 1.

Column 4 :

Col. 4 = (series C 115-132, Cols. 129 + 130 + 131) X Col. 1.

Column 5 :

Decade estimates of the annual wage income of common labor, Col. 4, Figure 4 discounted at 6% over a working life of 20 years.

Column 6 :

Col. 6 = Col. 5 X Col. 2 X 250%.

Column 7 :

Column 7 = Col. 5 X Col. 3 X 200%.

Column 8 :

Col. 8 = Col. 5 X Col. 4 .

Column 9 :

Col. 9 = Col. 6 + Col. 7 + Col. 8 .

TABLE 3.9. ANNUAL IMMIGRATION OF MECHANICS, ENGINEERS AND MANUFACTURERS TO THE U. S.: 1820-1855

Year ending Sept. 30	Mechanics	Engineers	Manufacturers
1820	269	12	13
1821	420	7	16
1822	283	16	18
1823	389	5	11
1824	289	20	10
1825	376	24	5
1826	593	14	16
1827	1,056	30	35
1828	1,334	33	29
1829	854	28	14
1830	942	37	8
1831	1,241	8	20
1832	4,145	35	16
Qtr. ending 12/31/1832	510	--	--
Yr. ending 12/31/1833	4,130	41	4
1834	3,642	60	40
1835	4,776	61	22
1836	7,838	14	--
1837	7,296	19	--
1838	4,636	13	5
1839	8,887	20	--
1840	9,474	40	--
1841	9,842	30	52

TABLE 3.9 - Cont.

Year ending Sept. 30,	Mechanics	Engineers	Manufacturers
1842	13,121	48	--
1st 3 Qtrs. 1843	5,155	26	13
1844	8,502	40	9
1845	9,836	53	13
1846	12,068	53	126
1847	24,567	35	503
1848	23,247	66	574
1849	29,564	142	382
1850	23,378	136	134
Qtr. ending 12/31/1850	5,131	25	27
Yr. ending Dec. 31			
1851	26,483	103	161
1852	24,120	91	122
1853	16,661	274	298
1854	31,470	213	364
1855	14,997	144	60
Totals	341,522	2,016	3,120

Source: William J. Bromwell, History of Immigration to the United States. Redfield, 1856, pp. 23-172.

market prices to homogeneous labor is a distortion of the theoretical meaning of factor intensity. Labor services should be measured in the same way as capital services, i.e. the labor requirement per unit of output measured in market prices. The process of capitalizing marginal products of the factors valued in market prices is a good proxy for the true measure of the factor's contribution to production in that the market price of the marginal product is a measurable unit in which to express, say, labor equivalents. Since we never encounter homogeneous labor it is necessary to measure labor input in "labor equivalents" to get the capital-labor ratio that is the one intended by theory. The estimates of human capital reduce non-homogeneous labor input to a common basis, the discounted value of the marginal contribution to output expressed in market prices. Then the "capital-labor" ratio or measure of factor intensity becomes the ratio of the input of equivalent capital services to the input of equivalent labor services.[49] Such a measure would represent the real factor proportions in production that is the common coin of economic analysis.

The evidence presented in this section demonstrated that it is particularly inappropriate to characterize American production as capital-intensive when the measure of intensity is dollars of capital per worker. Using this measure the long and fruitless search to understand the unique development of American technology has come to ground, since there is no reason to expect that American technology is more or less capital intensive than any other country's until evidence using the correct measure of factor proportions in production has been adduced. The comparison of the magnitudes of human capital inputs from immigration alone to gross physical capital formation raises more than a reasonable doubt that American technology was as capital intensive (on the correct measure) before 1860 as current discussion of technological development would lead one to suppose. It could well be the case that the enormous injection of skilled workers into the production process before 1860 served to bias the naive measure upward, hide the underlying reality that a unit of American output used as many or more labor equivalent services than capital equivalent services as say, Britain, and obfuscate our understanding of technological development; particularly when the theory and the empirical concepts with which we have conducted our analyses in the past have been so disparate.

The implications for a comparative study of factor intensities, measured in correspondence with the theoretical concept, will not be pursued further here.

CHAPTER IV

AN HISTORICAL FRAMEWORK FOR THE STUDY OF AMERICAN TECHNOLOGICAL DEVELOPMENT

Starting with some rather general propositions relating to questions of technical knowledge, technological choice and embodied technology, a set of paradigms for the historical analysis of technological change will be developed in this chapter. These models, which are developed with a view toward operationality in the data scarce world of the early nineteenth century, will be useful in future work to draw inferences from available empirical data.

Technological Choice: The Appropriate Unit of Analysis

It would appear to be desirable to define clearly and limit rather strictly the type of production function used in this type of inquiry. If one were to lean to traditional approaches to the historical study of technology he could choose either of two distinct methodologies. The first, employed by traditional historians or historians of technology, would be to taxonomize technical change as a sequence of improvements related to a specific machine or process. The second approach is to formulate the problem of "technical choice" in terms of economy wide behavior using the most general or mathematically tractable production function available. Each of these approaches suffers a crucial limitation. In the first case it would be the inability to carry the taxonomic approach to higher levels of generalization that would sharply limit its usefulness for purposes of historical interpretation. The latter approach is limited by the impossibility of dis-aggregating economy-wide behavior into historically comprehendable units of analysis, i.e., firms or industries. For example if the approaches used by Temin and Fogel could not explain the behavior of any single firm or industry, in what way would economy-wide behavior be "explained." The Temin-Fogel or aggregative approach to the use of production functions for historical analysis bases its predictive power on a derived relationship, i.e. the underlying industrial structure, inter-relationships, and technology itself. As Dusenberry has pointed out, derived relationships lose their power to predict when the fundamental, structural and behavioral relationships on which they are based change. From all accounts the fundamental conditions of production were constantly being modified throughout the first half of the nineteenth century if for no other reason than technical progress itself. To then use a stable, aggregative model based on the assumption of a static underlying structure as an exploratory tool in a dynamic environment would be logically inappropriate, particularly if one were trying to explain technological change by employing a model that presumed its constancy. The use of the stability inherent in aggregate relationships over relatively long periods of time to develop explanations of the causes of change in the micro units of analysis in particularistic settings is unlikely to predict this type of economic behavior accurately, except by chance. Further, each of these approaches, traditional and aggregative, uses a unit of analysis that is not coincident with the pattern of ownership and decision making. If technological change represents the results of conscious and deliberate action on the part of entrepreneurs and engineers, then the problem is best studied in a frame of reference that corresponds to the economic and organizational boundaries of decision making activity. In short, the current approaches to the study of technology are each invested with an irreversability of logic that prevents the unit of study from coinciding with the scope of decision making and control thus severing the logical connection between entrepreneurial choice and its manifestation in changing technologies.

The Replacement Decision and the Adoption of New Technologies

In Chapter II it was shown that when techniques are small in number and discrete the elasticity of substitution will not be a continuous and responsive function of relative factor prices. Factor substituability arose as a linear possibility in the production function defined as follows:

$$Q = \sum_i A_i \min (K_i, L_i)$$

where A_i is the proportion of industry (or firm) output produced by the ith fixed coefficient technique. As the A_i's change the observed capital-labor ratio for the firm or industry will also change. If substitution means something more than just a linear possibility, then as the price of labor rises relative to the price of capital, the percentage of industry (firm) output produced by more capital intensive techniques should also rise. This change in factor proportions implies a decision either to replace an existing machine, process or plant at the level of the firm, or in the case of industries, the replacement of whole firms or the entry of new firms opting for production facilities that are relatively capital intensive.

In analyzing embodied technological change it must be recognized that the choice of techniques is part and parcel of the decision to replace existing equipment. These future choices are circumscribed by past choices which are embodied in the existing production facility.[1] There are essentially two versions of the production function: the first is simply a matrix of developed machines or processes, the second is the set of possible techniques that could exist from exhaustive combination of all scientific knowledge. The latter version is appropriate for long-run analysis because the influence of relative factor prices does not affect the arguments of this function, rather they are defined by technical possibilities only. With respect to the short-run, or what shall be termed the operational production function, relative prices intrude at several decision making levels. First, what to design and how to design it are influenced by relative prices. Thus the range of machines offered for sale to firms is influenced by relative costs impinging on the design process. Only after this step does the process typically analyzed by economic theory come into play, i.e. the selection of specific equipments from the range offered by machine producers. This bifurcation of the short run or operational function raises the possibility that the sets of relative prices affecting designers and ultimate users may not be equal nor in fact pertain to the same factors and commodities. For example, the machine designer will consider the relative costs of materials comprising the machine itself and the wage costs of the various types of machine builders (founders and forgers vs. machinists), but the desideratum of machine design has always been efficiency defined as unit output per unit input with the variables having technical characteristics not money values, e.g. BTU's of input to BTU's of output. On the other hand the machine user considers the cost of the machine itself and the prices of the labor and materials that comprise the production process defined by the machine.

Thus at the level of the firm the choice of techniques is not only limited by the existing capital stock, but it is also circumscribed by the range of alternatives made available by machine builders. Then technology embodied in machines on the factory floor must pass through two decision processes each having its own set of relative prices. Past discussion of the labor scarcity hypothesis have considered only the relative prices facing the machine buyer or user. What is equally important is the relative costs facing machine designers who create, in the first instance, the possibility set or operational production function facing the machine users.

Having denied the labor scarcity hypotheses in the preceding chapter one is brought dangerously close to the position that "relative prices don't count." However, in terms of explaining technological development in the Pre-Crystal Palace, American economy the above argument simply says that we may have been looking at the wrong (or an incomplete) set of relative prices. If then American technology diverged from the British model, the relative costs affecting the process of machine design and construction may have had as much or more influence on technological development as the relative costs of capital and labor facing the user, and further these relative costs are not the same within a given economy.

Holding in abeyance the empirical question of the relative design costs confronting American machine builders, the analysis here will concentrate on showing how the linkage of the economic present to the past through investment, and the alteration of the simple, marginal rate of technical substitution of labor for capital by the presence or absence of third, fourth, . . . n^{th} factors in the production (cost) function influences the choice of embodied techniques through the replacement decision.

The replacement decision is made by comparing the operating costs of the present process with the total costs (operating plus capital) of the new process. If the operating costs of the old process exceed the total costs of the new, then replacement is profitable and should be undertaken. When these two costs are equal the firm is indifferent between adopting the new equipment or retaining the old. By equating the costs, the factors affecting differential wage costs between two processes can be analyzed to determine how sensitive the replacement decision is to differences in this element of cost. The controlling condition is,

$$(OC)_{old} = (OC+CC)_{new}$$

where

OC = operating costs

CC = capital cost

This condition could be portrayed as follows:

$$(1) \qquad V(W_oL_o+m_oM_o+s_oS_o+T_o) = C(d+i+r) + V(W_1L_1+m_1M_1+s_1S_1+T_1)$$

where,

V = volume of output in units per year

W_o = wage rate with the present process per unit of output

L_o = labor services with the present process per unit of output

m_o = material price with the present process per unit of output

M_o = material input with the present process per unit of output

s_o = average cost of scrap and rework with the present process per unit of final output

S_o = units of scrap and rework with the present process per unit of final output

T_o = unit value of time to the firm, per unit of output

C = purchase price of the new equipment

d = depreciation charge

i = interest rate

r = risk premium

The operating cost parameters and variables on the right hand side of equation (1) denoted by subscript 1 are values of the defined symbols for the new process. Equation (1) can be expressed in two forms for our purposes.

$$(2) \qquad \Delta wL = [\frac{C'}{V} - \Delta mM - \Delta sS - \Delta T]$$

or

$$(3) \qquad V = \frac{C'}{(\Delta wL + \Delta mM + \Delta sS + \Delta T)}$$

where,

$$\Delta wL = W_oL_o - W_1L_1$$
$$\Delta mM = m_oM_o - M_1M_1$$
$$\Delta sS = s_oS_o - s_1S_1$$
$$\Delta T = T_o - T_1$$
$$C' = C(d+i+r)$$

In equation (3) ΔwL may be interpreted as the difference in labor costs between the present and the new process that can exist before replacement will take place. The term C'/V is the user cost per unit of output of the new equipment. If this term is large it will increase ΔwL and widen the gap in labor costs tending to make replacement less likely if only small differences in wages or unit labor requirement exist between the two competing processes. The other terms in equation (2) ΔmM, the difference in unit material costs; ΔsS, the difference in unit scrap and rework costs; and ΔT, the difference in the time variable will all reduce the required difference in wage costs to bring about replacement if they are large and positive as defined. That is the new process must save material, reduce scrap and rework and reduce through-put in the production system for a given level of user cost per unit of output if the firm is to replace an existing process, when the potential savings in wage costs are relatively small.

For example, if the new process uses more of the same kind of material per unit of output as the existing process ($M_1 > M_o$), or if it uses a more expensive material ($m_1 > m_o$) then the wage cost differential will be widened. If the new process is apt to produce more scrap per unit of final output ($S_1 > S_o$) or if unit scrap values rise ($s_1 > s_o$), the wage costs differential necessary to bring about replacement will be likewise increased. Unit scrap values depend on material costs, labor costs, and the number of operations preceding the occurrence of the defect. If the new process tended to produce defects after the 20th operation versus the 5th operation in the old then the unit value of scrap would rise for the new process, ceteris paribus.

Once replacement has taken place the relative capital intensity of the new process will depend upon whether $[C_o'/L_o \lessgtr C_1'/L_1]$, these crude proxies for capital intensity will in turn depend upon the intervening magnitudes ΔmM, ΔsS and ΔT that serve to loosen the simple connection between the marginal rate of technical substitution of labor for capital and relative factor prices. In order for an existing process to be replaced, the new techniques must reduce cost in a number of ways some of which are not directly related to pure wage costs alone. That is, as we concentrate on the firm as the unit of analysis, operating costs as perceived at this decision making level include more elements than have been specified in simple substitution models. While the specification giving rise to equation (2) is by no means exhaustive it does portray an essential portion of reality and is, in principle, operational.[2]

Equation (3) shows the annual volume of production required given user cost per annum, and the other cost element differentials. It is similar in concept to Paul David's threshold model.[3]

However, the relative prices in equation (3) that govern the replacement decision, particularly with respect to industrial processes, are more inclusive than in David's model being:

$$\left[\frac{(d+i+r)}{(\Delta w+\Delta m+\Delta s+\Delta t)}\right]$$

instead of,

$$\left(\frac{d+.5i}{w}\right)$$

David does not include the possibility of wage rate changes as a result of adoption of the new technique, and erroneously includes only one-half the annual market rate of interest as the opportunity cost of holding the fixed capital associated with the new process.

Time in Production and Measures of Improvement

To this point very little has been said about the variable T, the unit value of time to the firm per unit of final output. While the production function of economic theory includes units of the variables per unit of time; time itself is not recognized explicitly as a factor of production or element of cost. Looking at costs to the firm for a moment, it will be recognized that time does have value in production (and other activities as well) which may be viewed in the nature of an opportunity cost. If the adoption of a new machine, process or plant alters the total time of production from receipt of raw material through final inspection, which can be thought of as total system through-put time; reduces work in process or final goods inventories; or reduces idle time due to interferences in serially interdependent productive units then, the value of the change in these magnitudes per unit of final output is the unit value of time to the firm in production. All of these, elements are combined in the production function in the definition of the variable output per unit of time. In treating productivity growth within the framework of a standard production function, technological change is defined as obtaining the same output with less real input or more output with the same input. However, the aggregative nature of such functions suppresses the precise nature of these time related characteristics of technical advance. Once a firm has decided to produce, say, V units per year then it will be worth something to be able to do so in the shortest possible production period. The shortening of the production period creates opportunities for reducing the costs associated with idle factors (labor), as well as, semi-fixed charges; and is desirable if for no other reason than the shifting of inventory costs forward to the channels of distribution.

Any production process can be characterized in one of two ways either as an assembly line (A-L) where there is a strict priority in the sequence of operations, or as a job shop (J-S) where the work may flow through the production facility in any one of a large number of feasible sequences with minimal priority constraints.

An assembly line could be characterized as:

h_1 station (1) h_j station (j) h_n station (n)

where each station is described in terms of the mean and standard deviation of its own production time (h, σ). Let j denote a station and k an observation. Then

$$\overline{H} = \sum_j \sum_k h_j k_k$$

However, $\overline{H}$ the mean system through-put will be assumed normally distributed with

$$\mathrm{var}(\sum h_{jk}) = \sum \mathrm{var}(h_{jk}) + \sum_{j_i \neq j_2} \mathrm{cov}\ h_{j1}h_{j2}$$

$$\mathrm{cov}(h_{j_1} h_{j_2}) = \sum_{k} \sum_{j_1 \neq j_2} (h_{j_1,k} - \overline{h}_{j_1,k})(h_{j_2,k} - \overline{h}_{j_2,k})$$

if the station times (h's) for processes j_1 and j_2 are independent, cov $(h_{j_1} h_{j_2}) = 0$.

Thus total system through-put, H, can be estimated from the (h_j, σ_j) of each station comprising the A-L production system. The expectation of H will be $\overline{H}$. However, $\overline{H}$ will be a stochastic random variable, and it is the variance of $\overline{H}$ that is of particular interest. Technical change may not only take the form of reducing $\overline{H}$, but also minimize var(H) by diminishing some of the individual $\mathrm{var}(h_j)$'s or by reducing interdependency among the station times $\mathrm{cov}(h_{j_1} h_{j_2})$. In fact, we would expect those stations with large $\mathrm{var}(h_j)$ to be the most likely areas for technical advance. Technological change not only implies the reduction of mean station times, but also station and over-all system variability.

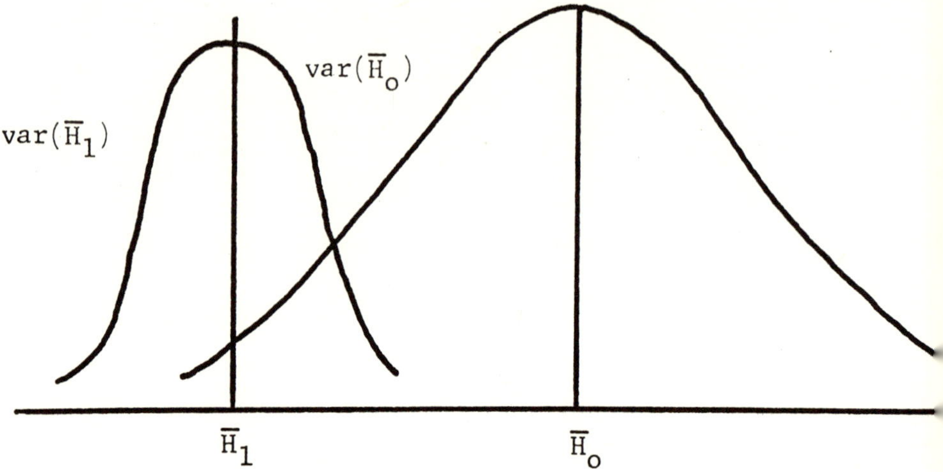

FIGURE 4.1

THROUGH-PUT VARIABILITY FOR TWO HYPOTHETICAL PRODUCTION PROCESSES

In Figure 4.1 we see that $\overline{H}_1 < \overline{H}_o$, i.e. expected system through-put has been reduced. This is the common notion of technological change. We could say the technological change had occurred unequivocally, however, if and only if $\overline{H}_1 < \overline{H}_o$ and $\text{var}(\overline{H}_1) < \text{var}(\overline{H}_o)$. It would also be possible to say that technological change had improved productive efficiency from the standpoint of the entire system even though $\overline{H}_1 > \overline{H}_o$, if $\text{var}(\overline{H}_1) < \text{var}(\overline{H}_o)$ i.e. if the variability of through-put time in the new system was significantly less than with the initial technique.

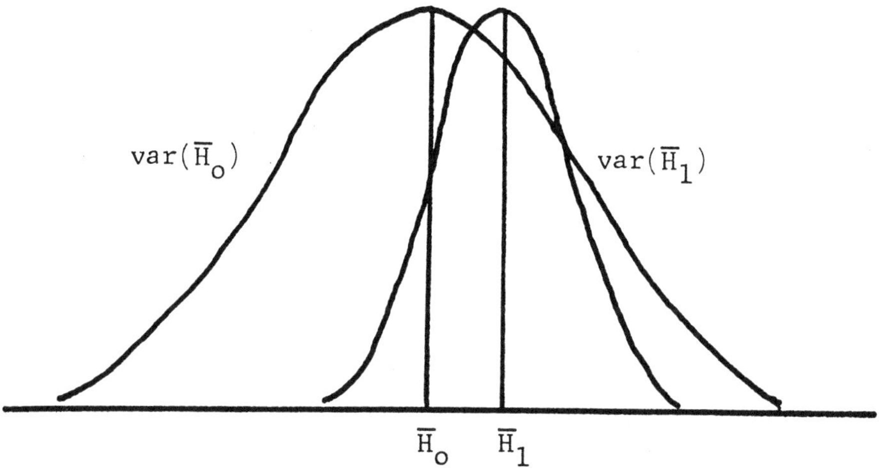

FIGURE 4.2

MEAN THROUGH-PUT AND PROCESS VARIABILITY AS MEASURES OF TECHNOLOGICAL CHANGE

In Figure 4.2 the determination of whether or not positive technological change had occurred would depend on the relative magnitude of the differences between $\bar{H}_o$ and $\bar{H}_1$ and var($\bar{H}_o$) and var($\bar{H}_1$). That is technological improvement could exist even if the expected system through-put, $\bar{H}_1$, had increased, providing system variability was reduced sufficiently.

How Time Enters into the Production System

This point can be seen in an intuitive way by considering a hypothetical example using a "clock" that records the time a work piece arrives at a given station and the time it leaves. This technique is commonly employed in queuing theory for a class of problems known as waiting lines.

In Table 4.1 the first unit or workpiece arrives at machine 1 when the clock reads zero hours-zero minutes (0:00). The processing takes two minutes, hence it leaves machine 1 when the clock reads (0:02), and is assumed to arrive at machine 2 instantaneously. In similar fashion the first work piece passes from machine 2 to machine 3. However, after the first workpiece leaves machine 1, that machine can accept a second workpiece. If the first machine finishes the second workpiece in two minutes the clock will read (0:04) when that piece is ready for machine 2. However, machine 2 is still processing the first piece, unit 1 does not leave machine 2 until (0:05), and a one minute delay occurs between the first and second machine on unit 2. Such delays arise even if h_1, h_2 and h_3 are

TABLE 4.1. ARRIVAL TIMES, PROCESS TIMES AND DELAYS FOR A 3-STATION PROCESS

Unit	Machine 1		Delay	Machine 2		Delay	Machine 3	
	Arrive	Leave	1-2	Arrive	Leave	2-3	Arrive	Leave
1	0:00	0:02	0	0:02	0:05	0	0:05	0:07
							(1 minute idle)	
2	0:02	0:04	1	0:05	0:08	0	0:08	0:12
3	0:04	0:06	2	0:08	0:11	1	0:12	0:15

$\bar{h}_1 = 2$ min. $\bar{h}_2 = 3$ min. $\bar{h}_3 = 3$ min.

constants. These delays may be minimized by a technique known as line balancing, however line balancing requires the use of multiple processes at some stations, e.g. 2 machines at step 2, and in some cases the capacity required to satisfy the rule of multiples may exceed the desired volume of production of the firm. If the h's are random variables that are distributed with a large dispersion, then even line balancing cannot eliminate delays and idle time. Only the conscious reduction of machine of process variability can reduce lost time and minimize system through-put.

The alternative formulation of the organization of production treats the facility as a job shop with each machine capable of performing operations on more than one part or "job." In order to appreciate the complexity of sequencing, consider the case in which four jobs must be done, each requiring time on five machines. There can be $(4!)^5$ or 7,962,624 different sequences, some of which, however, may not be feasible due to the fact that some of the required operations must be performed in a specified order. These "precedence requirements" arise from the technology of the manufacturing process. (A hole must be reamed before it can be threaded, etc.).

To illustrate the sequencing technique briefly, consider that machine times for four different operations to be performed on each of two jobs are as follows:

TABLE 4.2. OPERATION TIMES FOR TWO JOBS

MACHINE / JOB	a	b	c	d
1	4	8	10	2
2	4	10	6	12

Thus there are $(2!)^4$ or 16 possible sequences. Assuming that some of these sequences are technically impossible, it would be very desirable to eliminate the unfeasible sequences. Further, among the remaining feasible solutions, there are some which could not possibly be optimal and it would be desirable to identify these also. Akers and Friedman[4] have developed a technique for accomplishing such elimination which employs symbolic logic. This technique yields a subset of sequences one or more of which is optimal.

Suppose the required sequences of stations for each job are as follows:

job 1: a, b, c, d

job 2: d, b, a, c

Omitting the logical operations necessary to extract the possibly optimal, feasible programs, it can be shown that there are four programs which satisfy the necessary (but not necessarily sufficient) conditions for both technological feasibility and optimality. Each of the four programs can be displayed by a Gantt Chart as shown in Table 4.3.

TABLE 4.3. GANTT CHART FOR FOUR FEASIBLE AND POSSIBLY OPTIMAL PROGRAMS

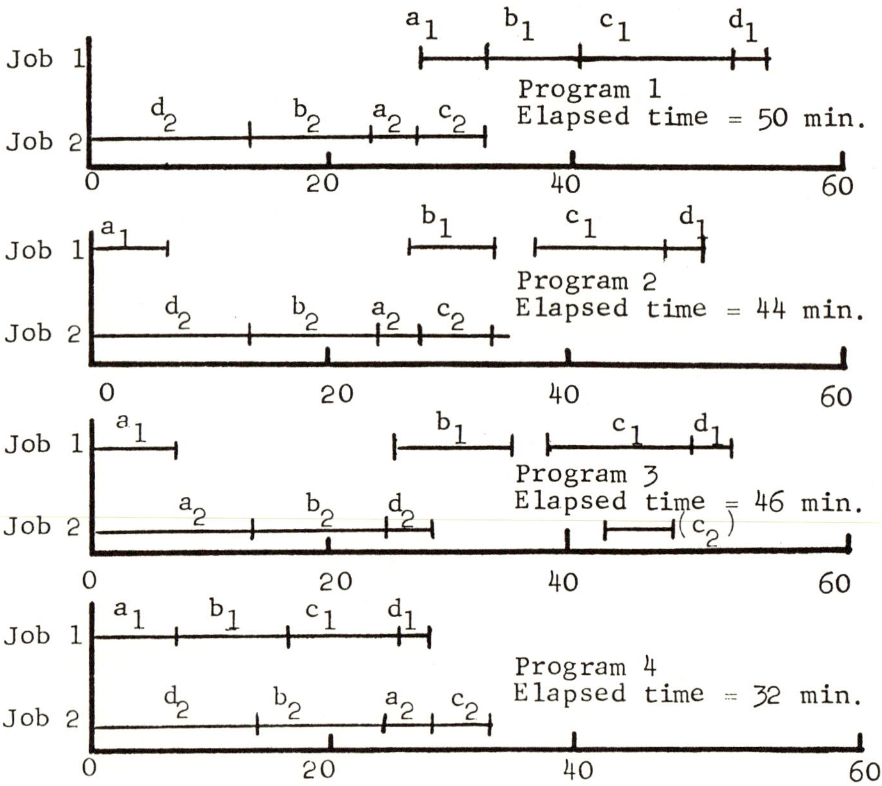

Analytical solution to more complex examples of m x n sequencing problems do not exist, for example when two or more machines of the same kind comprise a given work station, where alternative routing is allowed, when machines times are random variables or where machines are subject to down-time; system simulation is the only method of solution. The use of sequencing models and queuing illustrations for production systems characterized as job shops or assembly lines highlights the conclusion that system through-put is as much a function of organization and scheduling procedures (dis-embodied technology) as it is of individual station times (embodied technology).

The principle benefit to be derived from this view of production and the impact of technical advance is that in principle we have a method for separating the effects of embodied from dis-embodied technological change. If we could reproduce the routing sheets of a firm and gather data on process time requirements over a span of years it would be possible to quantify the effects of the two types of change. Say we have the route sheets, production scheduling or sequencing programs for a firm in 1850 and 1860 along with unit process times for those years and an idea of the precedence requirements involved in the operation. Then by simulation the 1860 output could be "produced" with an 1850 routing and 1860 station times to determine the magnitude of embodied technological change. Simulating 1860 production with 1860 routing and 1850 unit production times would measure the effect of dis-embodied technological change. If the calculation were properly constrained the two effects would sum to the total change in productivity between the two years.

The Supply of and Demand for New Machines

Returning now to the question of the relative costs facing machine builders i.e. supply costs of machines, a simple model will be developed to show the probable way in which American machine prices diverged from those of England. The analysis proceeds from the notion that there must be some price, determined by economic factors, that firms (users) will be just willing to pay to acquire a new machine or process i.e., invention. This willingness on the part of users to acquire the newest or best practice embodied technique may be viewed as the demand for invention facing machine builders. On the other side of the analysis is the willingness of builders to supply new machines (inventions), as governed by the relative prices impinging on their decision whether or not to devote resources to this end. Both relationships may be expressed in terms of the demand and supply of new machines. The demand for new designs embodied in machines can be written as,

$$D = f(p_o; w_1 \ldots w_m; m_1 \ldots m_n; d; r; i)$$

where,

p_o	=	price of final output
$w_1 \ldots w_m$	=	wages of the types of labor employed in machine using industries
$m_1 \ldots m_n$	=	prices of various material inputs to final product
i	=	interest rate
d	=	depreciation rate
r	=	risk premium (%)

The supply of embodied best practice, techniques is,

$$S = g(p_m; w_{m+1} \ldots w_q; m_{n+1} \ldots, m_t; i, T)$$

where,

p_m	=	price of machines
$w_{m+1} \ldots w_q$	=	wages of the types of labor employed in the machine building industries
$m_{n+1} \ldots m_t$	=	prices of various material inputs to machines

i = interest rate

T = technology

The point to note here is that in the specification of the demand and supply function for new machines, each function is determined by different sets of output prices, wage rates and unit material prices. Corresponding to each of these functions would be the marginal relationships derivable, in principle, from the respective total revenue and cost functions. The first derivative of the total revenue function will be called the marginal benefit function (MB) or the demand curve for new machines facing machine builders. It is called the marginal benefit function because for given levels of final output in the machine using industries it shows the incremental value of new machines to users at different rates of output of new machines. The first derivative of the total cost function facing machine builders is the marginal cost (MC) curve in that activity. Above the average variable cost curve, the MC curve and the supply curve of new machines coincide. According to the traditional version of the labor scarcity hypothesis the relative prices which determined the demand for machinery, and hence output, were those facing machine users and would correspond to the relative prices given in the derivation of the MB function. However, the conditions underlying the supply curve of new machines are never mentioned at all. Figure 4.3 shows the prediction of the labor scarcity hypotheses allowing relative prices to work only through the MB function while the condition of supply are assumed equivalent in England and America.

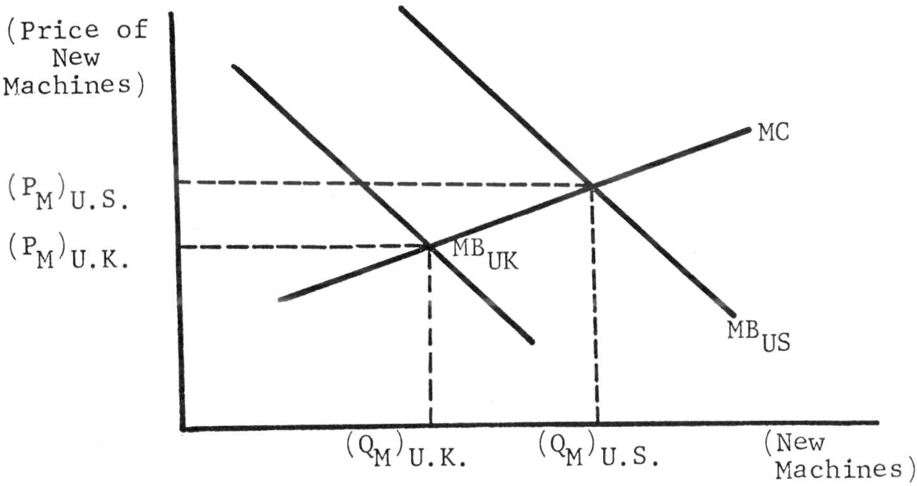

FIGURE 4.3

THE MODEL OF THE NAIVE LABOR SCARCITY HYPOTHESIS

In Figure 4.3 we see the price and output of new machinery in America both greater than in England due to the fact that the American MB function lies rightward of the England one. The reason for this could be given as the traditional ones, i.e. domestic manufactures sold at higher prices in America and labor was dearer relative to capital as well.

At the beginning of the nineteenth century the MC curve of England was probably well below that of America for the reason that English technology was more advanced and that she enjoyed an abundant supply of those skills and materials essential to machine building relative to America. Thus we find more new machines being produced in England, even though the American MB curve is rightward of England's, and very much cheaper than those made in America. Figure 4.4 shows the hypothetical, relative position of the curves of these two countries in 1800. Assuming that the relative position of the MB curves did not change by 1850, but that the influx of engineers and skilled workers into America was responsible for a downward shift in her supply curve of new machinery, then America would be producing more and cheaper new machines than England by this date. Figure 4.5 shows this condition.

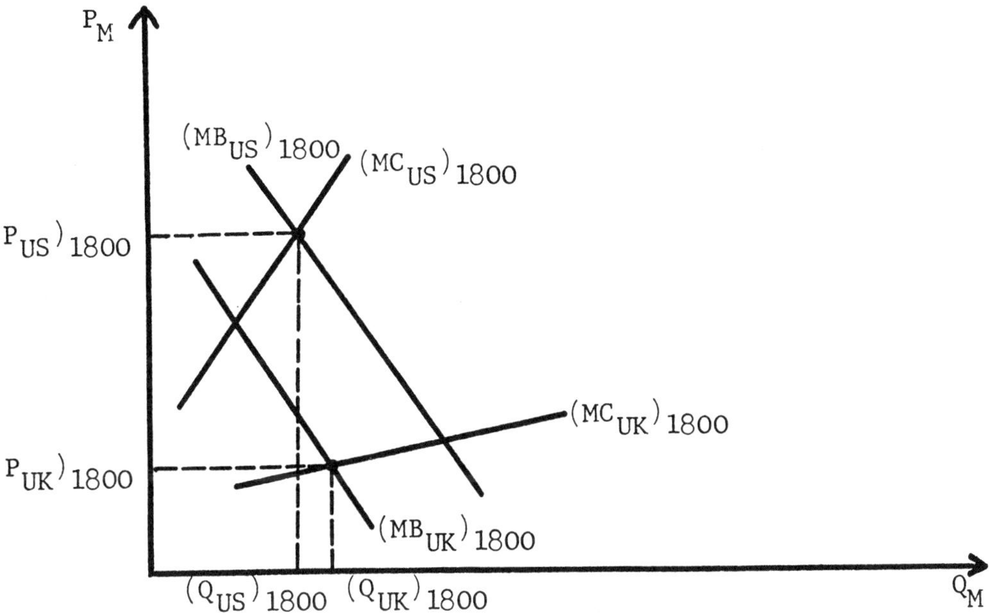

FIGURE 4.4

HYPOTHETICAL SUPPLY AND DEMAND CONDITIONS IN THE U. S. AND U. K. IN 1800

What is worth noting here is that this conclusion i.e. $(P_{UK})_{1850} > (P_{US})_{1850}$ and $(Q_{UK})_{1850} < (Q_{US})_{1850}$ could have been obtained even if the MB functions were identical. In other words, as long as the underlying economic factors were favorable to the change in relative position of the new machinery supply curves of the two countries from 1800 to 1850, we can account for the emergence of the United States as a technological leader and innovator without necessarily appealing to the

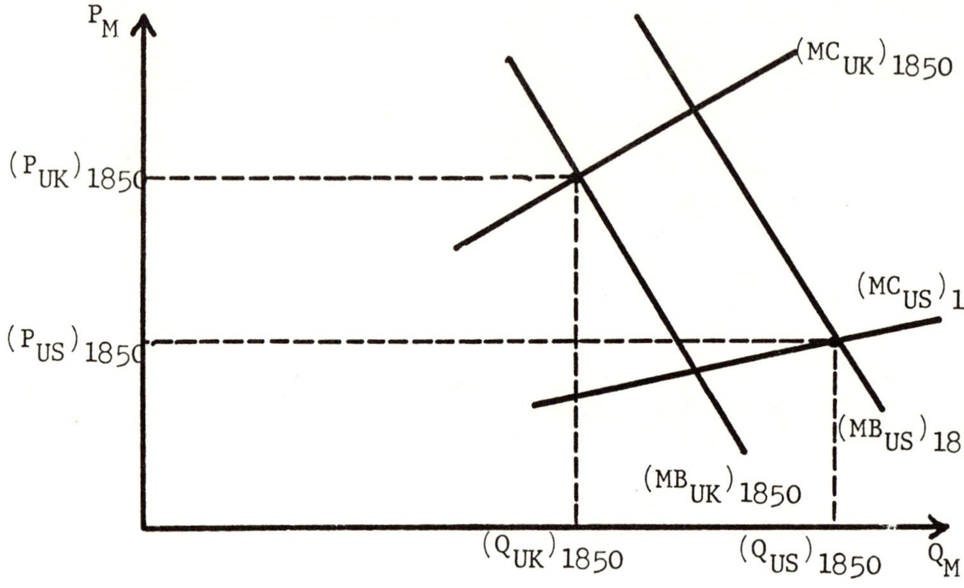

FIGURE 4.5

HYPOTHETICAL SUPPLY AND DEMAND CONDITIONS IN THE U. S. AND U. K. IN 1850

divergence of relative factor prices in the machine using industries, i.e. the traditional explicandum of the labor scarcity hypothesis. Thus the labor scarcity hypothesis, by the very nature of this implicit focus, i.e., machine users as opposed to machine builders, has led the inquiry as to the causes of technological divergence away from the most potentially fruitful area of inquiry, the cradle of embodied technology itself--the machine building industries. In planned later work this focus will be adopted. However, at present it might be well advised to explore how and why such marginal analysis could conceivably be used as a predictive tool in the economically imperfect world of the early nineteenth century.

Market Structure in Early Nineteenth Century America: The Impact on Factor Substitution

On the eve of the nineteenth century such American manufacturing as did exist was probably best characterized by the model of monopoly. These manufacturers, particularly in the metal working trades, were essentially regional or local monopolists in their product markets and near monopsonists in the factor market.[6] Another characteristic of their existence is to be found in the nature of the product and the conditions determining its demand curve. The products were basically utilitarian in nature and subject to a relatively inelastic demand until the local market had been saturated. Axes, shovels, saws, nails, and hand tools of various sorts are goods that every household in a given locality needed. After each household had obtained its requirements of these items, additional units could be sold only by the manufacturers' reducing price rather severely. Of course this rather static view abstracts from the formation of new households. But in the short run, the manufacturer faced a market of relatively fixed size and fixed total potential demand. The characteristics of this market are shown in Figure 4.6.

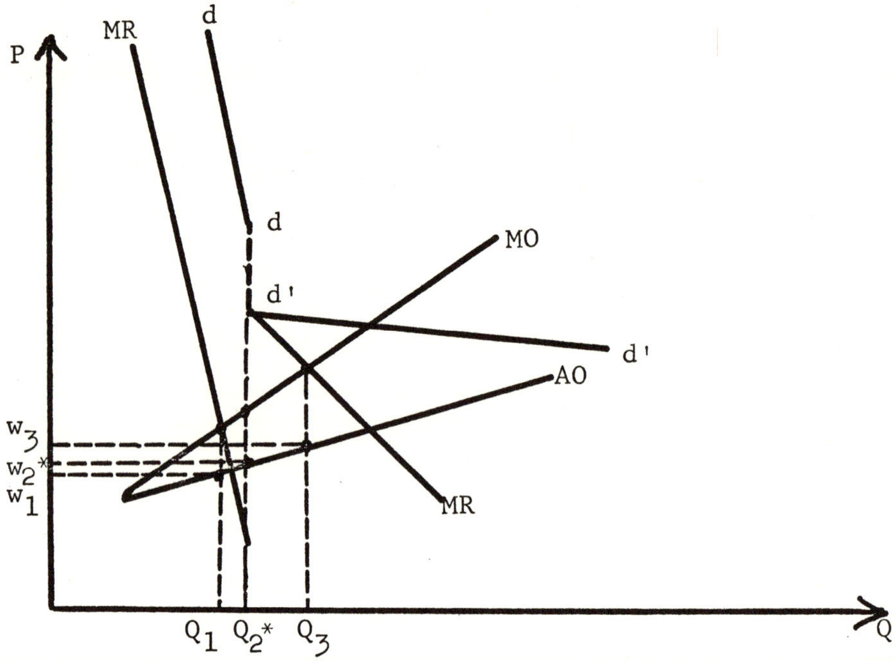

FIGURE 4.6

DEMAND AND SUPPLY CURVES FACING A LOCAL MONOPOLIST PRODUCING UTILITARIAN GOODS

Until Q* is reached the demand curve facing the manufacturer is given by d-d, after that point additional units can be sold in the local market only if prices are reduced drastically. After output Q* where every household has, say, one axe the demand curve, d'-d', becomes more elastic. Assuming the local monopolist is also a monopsonist in the labor market and that only wages enter into production costs, the marginal outlay curve (MO) will lie everywhere above the average outlay curve (AO) as output expands. Any one of three output points satisfies the condition that MO = MR, and gives rise to three possible ruling wage rates w_1, w_2* and w_3.[7] Thus the manufacturer would be indifferent among various wage rates on the range w_1-w_3, and this range shows his insensitivity to wage rate changes (increases) due to the particular behavior of local, utilitarian demand. Also his monopoly position creates positive profits at most feasible levels of output further reducing the pressure of high wages on his decision to alter factor proportions i.e. become more capital intensive. As long as positive profits exist there is no market pressure to alter factor proportions as a monopolist he may simply absorb the high wage costs, accept a somewhat reduced profit rate, forego the trouble of changing his production techniques and take out some of his windfall in the form of the quiet life. In this situation the local manufacturer is not necessarily concerned with wage costs to the exclusion of other capitalistic pre-occupations. One does not have to go as far as accepting the market model implied by Figure 4.6, the only necessary condition to obtain this conclusion is that the manufacturer be a monopolist in a local market. If this model is a fair characterization of the market conditions confronting American manufacturers at the beginning of the nineteenth century, the obvious question is why should it have been this way? The answer is to be found in the costs of transportation in this period. High transportation costs prevented these local monopolists from reaching out to each others markets. If unit transportation costs are included in the manufacturers average cost curve then he cannot sell in another market as long as,

$$MR_{rival} < [MC(T)]$$

where

MR_{rival} = marginal revenue obtainable from entry into a rival's market

$MC(T)$ = potential entrant's marginal cost defined to include cost of transport to rival market

Given this situation, a fall in transportation costs will permit previously local monopolists to invade, and in turn be invaded. Once this has occurred the analysis would proceed along the lines of the models of bilateral monopoly or oligopoly behavior. It is clear, however, that once unit transportation costs have fallen, limit pricing will no longer be able to forestall entry in and of itself if unit average costs are not in line with those of the potential entrant's. Thus the fall in transportation costs raises the threat of entry into local markets and puts general pressure on manufacturing costs and techniques, and brings into play the condition for the marginal analysis described previously. Table 4.4 shows typical freight rate ranges for the four modes of transport from 1780-1859 on a decadal basis. While these data are rather crude by themselves and mask over the classes of freight, geographic condition, and other special rate determining characteristics, they do present a rather clear picture of the trend of freight rates in the American economy to 1859. The process of the destruction of local monopolies most likely took place sometime after 1830 and before 1840. Table 4.4 only faintly supports this conjecture since we do not know how large a portion of the entrant's unit average cost was composed of freight charges, and how large a margin of safety unit freight costs provided for local monopolists. Of course, the bulkier and heavier the commodity in question, the more important the impact of transportation costs on this emerging process of competition. Another dimension of this emerging competition was the time it took to reach the desired destination, the irregular pattern of water transport in the North and the attendant uncertainties and high inventory costs. For example, in 1817 it took an average of 52 days to ship freight by keelboat, wagon and river from Cincinnati to New York City. Between 1843 and 1851 the shipment by a combination of steamboat, canal and railroad required 18-20 days. In 1852 the all water route took 18 days, while by rail the time was 6-8 days. In the 1850's a steamboat to New Orleans from Cincinnati and by packet from New Orleans to New York required 28 days.[8] The barriers to domestic competition and market development created by lack of adequate transportation in the second decade of the nineteenth century may be more clearly understood from these comments by the president and directors of the Chesapeake and Delaware Canal in 1816.

> A ton of goods by weight or measurement (of 40 cubic feet) was frequently brought from Europe, a distance of 3,000 miles, for 40s. sterling, or about $9; this rate admitted salt, coal,

TABLE 4.4. TYPICAL FREIGHT RATE RANGES (CENTS PER TON-MILE), 1780-1859

	Pack Horse or Wagon	River	Canal	Railroad
1780-89	30-90¢	1-13¢	- -	- -
1790-99	15-90¢	1-13¢	- -	- -
1800-09	13.5-70¢	5-10¢	5.5¢ (avg)	- -
1810-19	13.5-70¢ (33¢ avg.)	.5-6¢	1-8¢	- -
1820-29	12-30¢	- -	.5-8¢	3.5-13.5¢
1830-39	17¢ (turnpike)	- -	.5-8¢	3.5-11.4¢
1840-49	7-20¢	.2-1.5¢	.5-7.5¢	7.6 (avg)
1850-59	7-18¢	.2-1.5¢	.2-2.5¢	1.5-6¢

Sources: J. L. Ringwalt, Development of Transportation Systems in the United States, Philadelphia, 1888.

G. R. Taylor, The Transportation Revolution, 1815-60, Harper and Row, 1951.

L. Hunter, Steamboats on the Western Rivers, Harvard University Press, 1949.

J. M. Swank, Progressive Pennsylvania, Philadelphia, J. B. Lippincott, 1908.

C. E. MacGill, History of Transportation in the United States before 1860, Carnegie Institution of Washington, 1948.

> stone, lumber, and most bulky articles, which were imported from beyond the seas cheaper than they could be conveyed for a very short distance by land, since a ton of goods could not be carried on good roads for the same price more than 30 miles. The charges for conveyance through the United States were very great, even for fine and valuable articles; but on the bulky products of agriculture and on mineral productions they operated as a total exclusion . . . A coal-mine might therefore exist in the United States not more than 10 miles from valuable ores of iron and other materials, and both of them be useless until a canal were established between them, as the price of land carriage was too great to be borne by either. The conveyance of produce coastwise by sea did not lessen the expense in any manner proportioned to that of canals, or even of foreign vessels, for which the following reasons were assigned. First, the coasting trade was chiefly carried on in vessels which took their produce from and delivered it into small ports; these vessels were therefore generally small and were navigated at a far greater expense in proportion than large vessels. Second, the danger of coasting voyages was greater, which proportionally increased the risk and prices of insurance. Third, the delays of delivery, accident, etc., port charges, repairs, and other expenses were cumulative on small vessels; hence goods were frequently brought from Europe at nearly as low a cost as they could be conveyed by sea from Baltimore to Philadelphia, cities which by land were not more than 100 miles from each other. This was remarkably the case with respect to coal, which was brought into most of our maritime cities from Liverpool and sold as cheap as it could be brought to them from the James River, not one-sixth of the distance.[9]

The McLane Report also provides some insight into the condition of American manufacturers in the early part of the nineteenth century. Table 4.5 shows sample rates of reported profit in New England in 1831. While Table 4.5 is not an exhaustive tabulation of the data contained in this two volume work, it represents the results of random samples drawn from eight industries in seven states, and as such is a reasonable approximation to the profitability of manufacturers in this period. While the accounting procedures generating these data were notoriously non-uniform, the profit rates are net of interest charges on borrowed capital, and most of them are net of depreciation. Most respondents indicated their capital cost was about six percent, with depreciation rates varying from 4 to 10 percent per annum. The opportunity cost of invested capital would have to be deducted from the profit rates reported in Table 4.5 in order to approximate the economic profit. If the relevant opportunity cost was the market rate, this deduction would be about six percent. Allowing for the deduction of this opportunity cost the economic profit reported by these firms in 1831 was positive and quite substantial. Since there are no observations for years prior to 1831 we cannot directly test Habakkuk's assertion that marginal profit rates were declining. However, since average economic profit in 1831 was positive and large it is doubtful that competitive

TABLE 4.5. SAMPLE RATES OF REPORTED PROFIT IN NEW ENGLAND IN 1831

Industry	Pop. Size	Sample Size	State	1831 Profit Rate	Previous Profit Rate	Comments
Cotton	280	17	Maine	1. 20%	Less	
			Massachusetts	1. 15%	10% 1830 0% 1829 Less 1828	
				2. 6%	15% 1830 15% 1830	Annual Net Profit: 6%
				3. "Less than 6%"	0% 1829	
			New Hampshire	1. 12%	--	
			Vermont	1. 6%	--	Average of 5 firms for three years
			Rhode Island	1. 7.75% 2. 6% 3. 15% 4. 10%	Until 1824: 0%	
				5. "No prof-its can be made." 6. 7% avg.		

TABLE 4.5. (Continued)

Industry	Pop. Size	Sample Size	State	1831 Profit Rate	Previous Profit Rate	Comments
Cotton (Cont.)			Connecticut	1. 3%	Less before 1831	
				2. 5.5%		Average since 1822
				3. 4.9%		1828-1831
				4. 4% on borrowed 10% on other		Failed repeated- 16 before 1831
Iron	68	8	Maine	1. 30%		No profit rate is given in a usable form for the state of Mass. Capital invested is reported but no earnings figures
			New Hampshire	1. 14%		on iron castings
				20%		Bar iron
				2. 22.5%		Iron Forge
				3. 25%		" "
				4. 6%		" "
				5. 18%		" "

TABLE 4.5. (Continued)

Industry	Pop. Size	Sample Size	State	1831 Profit Rate	Previous Profit Rate	Comments
Iron (Cont.)				6. 12% 7. 13%		Iron Forge Iron Furnace It is noted that the iron of New Hampshire is of superior quality.
			Vermont	1. 8% 2. 12% 3. 13% 4. 7% 5. 40%		
			Connecticut	1. 17.25%	$5-6,000 loss '26-'29 16.75% '30	
				2. 10% 16%		borrowed money non-borrowed money Reason for increase is that quality became better known.

Industry	Pop. Size	Sample Size	State	1831 Profit Rate	Previous Profit Rate	Comments
Marble	5	5	Maine	1. 10% 2. 10% 3. 8% 4. 8%		
			Vermont	1. 7.5%		
Paper	28	4	Massachusetts	1. 16% 2. See Comm.		For one county where many of the paper plants were located the response was "no definite answer. Most said they made little or nothing."
			Vermont	1. 23%		
			Connecticut	1. 6%		
Leather	213	12	Vermont	1. 12% 2. 33% 3. 26% 4. 10-1/2% 5. 10-1/2% 6. 25%		

TABLE 4.5. (Continued)

Industry	Pop. Size	Sample Size	State	1831 Profit Rate	Previous Profit Rate	Comments
Leather (Cont.)				7. 14% 8. 25% 9. 12%		
			Connecticut	1. 8% 2. 8% 3. 6%	8% previous years	
Wool	71	12	Maine	1. 12%		
			Massachusetts	1. 10-12% 2. 10%		
			New Hampshire	1. 3%		This is a combination of a cotton and a wool factory.
			Vermont	1. 12% 2. 8% 3. 6%		Wool was often manufactured in the home for home use. One statistic cited in N.H. showed a loss of 17.75% for this type of home manufacture.
			Rhode Island	1. 6% 2. Less than 2%		

TABLE 4.5. (Continued)

Industry	Pop. Size	Sample Size	State	1831 Profit Rate	Previous Profit Rate	Comments
Wool (Cont.)			Connecticut	1. 12% 2. 6% 3. 9%		
Soap and Candles	13	2	Massachusetts	1. -10% to -15%	Last 3 yrs. 10-15%	
			Vermont	1. 1-12%		Several different firms
Black-smith	40	9	New Hampshire	1. 10% 2. 20% 3. 10% 4. 100% 5. 150% 6. 75% 7. 50%		
			Maine	1. 12%		
			Connecticut	1. 6%		

Source: U. S. Serial Set, U. S. House Documents, The McLane Report, Serial CCXXII, 22nd Congress 1831-32, No. 308, Vol. 1.

pressures were "pressing against available technology," or significantly disturbing the quiet life of the local monopolist. There are several categories of response returned to the Secretary of the Treasury that are quite pertinent in this regard. A portion of the questionnaire required a response as to the extent of competition and the proportion of output consumed in local, out-of-state, and foreign markets. The picture that emerges from the McLane Report in this regard is that over three-fourths of the respondents had "no competition," and that most were still selling in local markets 20-30 miles from the point of production. Such competition as was reported was from abroad. Thus in 1831 the tariff and ocean freight rates were the margin of protection for domestic manufacturers, and the unit value of these two elements was the protective cushion for whatever inefficiencies or deficiencies were to be found in American manufacturing practice before economic profits would be erased. In 1831 markets for American manufacturers were predominantly local in character, and available evidence indicates that economic profits were still large and positive. Competitive pressures did not challenge American manufacturing practice or technologies until after this date. This conclusion is, of course, a general one, and the experience of a few industries may diverge enough to prove the rule.

The American Doctrine of Imperfectability

Finally, the role of a rather intangible element in American technological development will be taken up. Professor Eugene Ferguson has called this an "elusive attitude or point of view of American technologists before 1840 that would be most useful to isolate."[10] Professor Ferguson has termed this notion the American Doctrine of Imperfectability, and I feel this doctrine is an indispensable part of the character of American technological development. The records left by early mechanics, correspondence, patent specifications and various reports all bear witness to the unmistakable presence of this operative attitude. It was clearly recognized as early as 1820 that a machine part or musket component could never be made to conform exactly to a mean or nominal dimension. Two mechanics using handicraft methods of production could not produce parts that were identical even if filing jigs were used as in the case of Eli Whitney's production scheme. Just as in the earlier part of this chapter, where process times were seen to vary about some mean or central tendency, the dimension generating function of a machine is a random variable with some distribution about a mean or nominal dimension. That is a machine reaming one inch holes will produce holes that vary in actual measurement about the nominal dimension of one inch, or a slabbing mill that is set to remove .01 inch of metal will remove various thicknesses around the .01 dimension if for no other reason than the wear of the milling cutter. This phenomenon is the geometric analog of the natural time variation in production discussed earlier. The implicit hypothesis contained in Ferguson's Doctrine of American Imperfectability is that American mechanics acted upon this doctrine by employing a version of dimensional tolerancing, as is standard drafting practice today, and established inspection functions that accepted or rejected machine work based on the specified functional tolerances.

In contrast to this the English mechanics adopted a different point of view on the process of designing and producing machine parts. After Whitworth's visit to America in 1853 to observe American manufacturing practice firsthand he was questioned by a Parliamentary committee. The curious position taken by this distinguished English engineer was that no part could be used "as is" directly after machining if the results were to be perfect; hand work and fitting must intervene before assembly.

A Birmingham gun-maker was asked by the same committee: "Do you think with your knowledge of gun-making, that it is possible by any machinery, in any reasonable time, and at any reasonable expense, to make gun materials which shall fit into one another perfectly without a great deal of manual labor?" The gunmaker replied, "Certainly not, by no means, it is impossible."[11] This exchange points up the basic operative philosophy of the English mechanic which simply did not admit the possibility of setting out to make machine work anything less than perfect. This state of mind caused the English mechanics to regard American-fabricated machine work to be distinctly second rate. It is only fair to point out, however, that the Birmingham gun-maker was quite right in terms of his mechanical, as opposed to the philosophical or potential, necessity of hand work. There were more metal files, indicative of hand shaping and fitting of parts, in the tool rooms of the Springfield Armory in 1850 than is consistent with the notion that interchangeable manufacture based exclusively on the inherent accuracy of machining processes had been obtained by that date.[12]

The doctrine of mechanical imperfectability was probably operative from the time of Whitney and Perkins, but remained unarticulated until the work of Walter Shewhart in 1931 entitled Economic Control of Quality of Manufactured Products. Shewhart was the first to make explicit use of statistical inference for controlling production processes, although in 1922 G. S. Radford's Control of Quality in Manufacturing had taken the intuitive position that "the control of quality is the correct starting point for economy." The first statement that machining costs are a function of dimensional tolerances was made by M. F. Spotts in 1958 in the ASME Transactions. Spotts' idea was that nominal dimensions heretofore had been toleranced with little or no regard for the physical limits of variability of the process on which they were to be manufactured. That is, in going from a tolerance of $\pm$ .01 to one of $\pm$.001, the manufacturing cost will increase, while the configuration and mean dimensions of the part remain unchanged; and further these costs are related to the inherent physical variability of the manufacturing process.

This recognition and incorporation of the inherent variability of production processes into design and engineering practices was a uniquely American phenomenon. Although unarticulated for almost 125 years and not finally stated in explicit form until very recent times, the ideas and attitudes underlying Ferguson's Doctrine of American Imperfectability were an enormously potent factor in the development of American technology in the nineteenth century for in the heads of hands of mechanics and engineers these ideas were operationalized to the point where they found their ultimate embodiment in the technology itself.

In this chapter an operational framework for the study of technological development has been presented. The elements influencing the replacement decision, and hence the adoption of new techniques, have been discussed with particular emphasis on the way time explicitly enters into consideration. The illustrative models presented here have emphasized the relationship between time as a factor of production and dis-embodied technology. The decision to adopt a new embodied technology by user firms has been presented in an augmented version of the labor scarcity hypothesis. However, the neglect of the decisions and variables impinging on the supply of innovations has been remedied in this extended analysis. One result of this approach has been to shift the historical focus to the relative cost conditions in the machine building, as opposed to the machine using firms for the purpose of understanding why American and British manufacturing practice diverged rather sharply after the middle of the nineteenth century. The appropriateness of this analysis given market structures in the American economy in the early part of the nineteenth century is discussed with reference to available empirical evidence. Finally, the role of attitudes as translated into specific mechanical practices, and their probable impact on technological development, is presented in an historical perspective.

CHAPTER V

CONCLUSION

That the American economy had developed unique manufacturing methods by the middle of the nineteenth century is in little doubt. Contemporary accounts and most of the literature in economic history assign a crucial role to the labor scarcity hypothesis in accounting for this development. The idea that relative factor prices determine, virtually by themselves, the cost and character of technological change is firmly rooted in the literature of this subject going back to the earliest days of the nineteenth century. That such beliefs should have held sway for so long is all the more remarkable when one considers that there is no systematic or connected account of the development of the systems of interchangeable manufacture and high volume production in the first half of the nineteenth century. In this regard not only is *the* past not known, but we do not even have *a* past that is coherent and capable of withstanding even the most elementary logical assault as has been demonstrated. This aspect of our past, which bears vitally on how and why America grew, in a few brief decades after the Civil War, to a position of unrivaled wealth in all of recorded history has suffered from the studied neglect that is accorded the truly vital elements of national experience. Myths, apparent realities and vested interests in ideas all conspire to produce a past that is at best ennobling, often comforting, but never intellectually satisfying. Such has been the historiography of this subject.

The preceding chapters have served as a preliminary excursion into the theoretical and emperical background conditions of American technological development. The limitations of previous attempts to develop a frame of reference or intellectual scaffolding with which to study the phenomena of interest have been analyzed. This step is essential if future work is to be directed away from avenues of research which are certain to be fruitless.

While portraying the broad economic conditions that operated in nineteenth century America and their relationships to technological change either as vehicles for testing existing theories or, in the case of immigration, an enabling condition for capital formation, the main thrust of this work has been to provide new perspectives for the study of our technological past in the hope that they will aid future research in developing a richer understanding of this aspect of human experience.

APPENDIX TO CHAPTER I

The factor-price-frontier for a Cobb-Douglas production function may be derived as follows. Let

(a) $$Q_M = GL^{a_1}K^{a_2} \qquad (a_1+a_2) = 1$$

(b) $$P_m \cdot Q_M = P_M GL^{a_1}K^{a_2}$$

by VMP = factor price

(c) $$\frac{\partial(P_M Q_M)}{\partial L} = P_M G a_1 \frac{L^{a_1}}{L} K^{a_2} = w$$

(d) $$\frac{\partial(P_M \cdot Q_M)}{\partial K} = P_M G a_2 L^{a_1} \frac{K^{a_2}}{K} = r$$

(e) $$P_M G a_1 L^{-a_2} K^{a_2} = w$$

(f) $$P_M G a_1 (K/L)^{a_2} = w$$

(g) $$P_M G a_2 L^{a_1} K^{-a_1} = r$$

(h) $$P_M G a_2 (L/K)^{a_1} = r$$

(i) $$(P_M G a_1)^{\frac{1}{a_2}} (K/L) = w^{\frac{1}{a_2}}$$

(j) $$(P_M G a_2)^{\frac{1}{a_1}} (L/K) = r^{\frac{1}{a_1}}$$

(k) $$= (i) \cdot (j)(P_M G a_1)^{\frac{1}{a_2}} \cdot (P_M G a_2)^{\frac{1}{a_1}}$$

$$= w^{1/a_2} r^{1/a_1}$$

or,

$$(P_M G)^{\frac{1}{a_2} + \frac{1}{a_1}} \left(a_1^{\frac{1}{a_2}} a_2^{\frac{1}{a_1}} \right) w^{-\frac{1}{a_2}} = r^{\frac{1}{a_1}}$$

simplifying further,

$$(P_M G)^{\frac{a_1}{a_2} + 1} \left(a_1^{\frac{a_1}{a_2}} a_2 \right) w^{-\frac{a_1}{a_2}} = r$$

$$(P_M G)^{\frac{1}{a_2}} w^{-\frac{a_1}{a_2}} \left(a_1^{\frac{a_1}{a_2}} a_2 \right) = r$$

$$(1) \qquad r = P_M^{\frac{1}{a_2}} G^{\frac{1}{a_2}} w^{-a_1/a_2} \left(a_1^{a_1/a_2} a_2 \right)$$

APPENDIX I TO CHAPTER 2

(1) let $Q = L^a K^b$; $a + b = 1$ see Note 1.

(2) then $r = \frac{w}{k} = \frac{P \cdot L}{P \cdot K} = \frac{-dK}{dL}$ in equilibrium

(3) $K/L = r \cdot \frac{b}{a}$, by $\frac{k \cdot K}{w \cdot L} = b/a = \frac{\text{capital's share}}{\text{labor's share}}$

(4) $d(K/L) = rd(b/a) + b/a \cdot dr$

(5) by definition

$$\sigma = \frac{d(K/L)/K/L}{\frac{dr}{r}} = \frac{[rd(b/a) + b/a \cdot dr]a/b \cdot \frac{1}{r}}{\frac{dr}{r}}$$

(6) $$\sigma = \frac{dr + r \cdot d(b/a) \cdot a/b}{dr}$$

(7) $$\frac{(\sigma - 1)dr}{r} = a/b \; d(b/a)$$

$$\frac{(\sigma - 1)dr}{r} = \frac{a\,db - b\,da}{a^2}$$

$$\left[\frac{b\;a(\sigma - 1)dr}{r}\right] = a\,db - b\,da$$

Note 1. The use of a Cobb-Douglas function may be objected to on the grounds of the well-known property that such a function always has unitary elasticity of substitution providing technical progress is neutral. However, if we admit non-neutral technical progress then the elasticities of production a and b change relative to each other altering the marginal rate of substitution of labor for capital at each capital-labor ratio, i.e., allowing to vary over time. To see this consider the following, if

(i) $$\sigma = \frac{\frac{dz}{z}}{dr/r}$$

where

$$z = K/L \quad \text{and} \quad r = w/k$$

then

(ii) $$\int \frac{dr}{r} = \frac{1}{\sigma} \int \frac{dz}{z}$$

and,

$$\ln r = \frac{1}{\sigma} \ln z + \ln c$$

(iii) or

$$r = cz^{\frac{1}{\sigma}}, \quad c = \text{constant of integration}$$

from $r = a/b \cdot z$ in Eq. (3), Appendix I

or

$$r = \frac{a \cdot K}{b \cdot L}, \quad \text{and (iii) above}$$

(iv) $$\frac{a}{b} = c(K/L)^{\frac{1}{\sigma}}$$

Thus, if a and b are not constant, i.e., there is non-neutral technical progress in the last half of the nineteenth century, then may vary over time, and does not necessarily have to be unity at each (K/L) ratio. Further discussion of this point is contained in Murray Brown, <u>On the Theory and Measurement of Technological Change</u>. Cambridge University Press, 1966,

(8) Since, $a = 1 - b$; $da = -db$

and, $a\,db - b\,da = a\,db - bd(1-b)$

then, $\frac{b\ a(\sigma-1)dr}{r} = a\ db - b\ d(1-b)$

(9) $\frac{b\ a(\sigma-1)dr}{r} = (1-b)db + b\cdot db$

(10) $db = a\cdot b(\sigma-1)\frac{dr}{r}$

or

$-da = a\cdot b(\sigma-1)\frac{dr}{r}$

Chapter 3. The Cobb-Douglas function may be generalized by introducing an exponential term multiplicatively to yield variable elasticities of substitution, c.f. C. E. Ferguson and R. W. Pfouts, "Aggregate Production Functions and Relative Factor Shares," International Economic Review, Vol. 3, September, 1962. However, for our purposes as long as the production elasticities are permitted to vary, then relative income shares will change yielding estimates of "implied" elasticities, i.e., elasticities which have changed over some relevant period of time as the economy moves to successive equilibria.

APPENDIX II TO CHAPTER 2

(1) $$\sigma = \frac{\dfrac{d(K/L)}{K/L}}{\dfrac{d(w/k)}{\frac{w}{k}}} = \frac{\dfrac{L\ d\ K - K\ d\ L}{L\cdot K}}{\dfrac{k\ dw - w\ dk}{k\cdot w}}$$

(2) $$\sigma = \frac{k\cdot w\ L\ d\ K - k\cdot w\ K\ d\ L}{L\cdot K\ k\ dw - L\cdot K\ w\ dk}$$

(3) $$d\ L = \frac{\sigma(L\cdot K\ k\ dw - L\cdot K\ w\ dk) - k\cdot w\cdot L\cdot d\ K}{k\cdot w\cdot K}$$

(4) $$d\ L = \sigma\cdot L\cdot\frac{d(w/k)}{w/k} - \frac{k\cdot w\cdot L\cdot d\ K}{k\cdot w\cdot K}$$

(5) $$d\ L = \frac{k\cdot w\ [K\cdot L\cdot\sigma\cdot d(w/k) - L\cdot d\ K\cdot(k/w)]}{(w/k)\ w\cdot k\cdot K}$$

(6) $$d\ L = \frac{k\cdot w\cdot K\cdot L\cdot\sigma\cdot d(w/k) - k\cdot w\cdot L\cdot d\cdot K\cdot(w/k)\dfrac{\sigma\cdot d\frac{w}{k}}{\sigma\cdot d(w/k)}}{(w/k)\cdot w\cdot k\cdot K}$$

by (1), $$\frac{(w/k)}{\sigma\cdot d(w/k)} = \frac{K/L}{d(K/L)}$$

(7) $$d\ L = L\cdot\sigma\cdot\frac{d(w/k)}{w/k}\left[\frac{k\ w\ K - \dfrac{k\cdot w\ d\ K\cdot L\cdot K}{L\cdot d\ K - K\ d\ L}}{w\cdot k\cdot K}\right]$$

(8) $$d\,L = L\cdot\sigma\cdot\frac{d(w/k)}{w/k}\left[\frac{k\ w\ K - \dfrac{k\cdot w\ L\cdot K}{L - K\ dL/dK}}{w\cdot k\cdot K}\right]$$

by $w/k = -dK/dL$

(9) $$d\,L = L\cdot\sigma\cdot\frac{d(w/k)}{w/k}\left[\frac{k\cdot w\cdot K - \dfrac{k\cdot w\cdot L\cdot K}{L - K(k/w)}}{w\cdot k\cdot K}\right]$$

(10) $$d\cdot L = L\cdot\sigma\cdot\frac{d(w/k)}{w/k}\left[\frac{k\cdot w\cdot K - \dfrac{k\cdot w\cdot L\cdot K}{\dfrac{w\ L + k\cdot K}{w}}}{w\cdot k\cdot K}\right]$$

(11) $$d\,L = L\cdot\sigma\cdot\frac{d(w/k)}{w/k}\left[1 - \frac{L}{\dfrac{w\ L + k\ K}{w}}\right]$$

(12) $$d\,L = L\cdot\sigma\cdot\frac{d(w/k)}{w/k}\left[1 - \frac{w\ L}{w\ L + k\ K}\right]$$

or

$$d\,L = L\cdot\sigma\cdot\frac{d(w/k)}{(w/k)}\cdot\pi$$

where

$$\pi = \frac{k\cdot K}{k\cdot K + w\cdot L} \doteq \text{capital's share of output}$$

and,

(13) $$\frac{d\,L}{dt}\cdot\frac{1}{L} = \sigma\cdot\frac{d(w/k)}{dt}\cdot\left(\frac{k}{w}\right)$$

F O O T N O T E S

FOOTNOTES

Chapter I

[1]E. Rothbarth, "Causes of the Superior Efficiency of U.S.A. Industry as Compared with British Industry." Economic Journal LVI, September, 1946, pp. 383-90.

[2]H. J. Habakkuk, American and British Technology in the Nineteenth Century. Cambridge, England: University Press, (1962).

[3]Peter Temin, "Labor Scarcity and the Problem of American Industrial Efficiency in the 1850's," Journal of Economic History, Vol. XXVI, No. 3, Sept. 1966.

[4]Robert Fogel, "The Specification Problem in Economic History," Journal of Economic History, Vol. XXVII, No. 3, Sept. 1967.

[5]Temin, p. 295.

[6]Temin, p. 291.

[7]Sidney Homer, A History of Interest Rates, New Brunswick, N. J.: Rutgers University Press, (1963).

[8]Fogel, pp. 303-304.

[9]Fogel, p. 307 and R. W. Fogel, "The Specification Problem in Economic History: A Correction," Journal of Economic History, Vol. XXVII, March, 1968, p. 126 has inadvertently written the bracketed term as $(P_M/\overline{P})^{a_1/a_2}$, his equation 13. Under the assumption

that a_1 and a_2 are equal in Britain and America, this does not alter his conclusion, however.

[10]The fact that the British and American economies were structurally different throughout most of the nineteenth century $(k_1)_{U.S.} \neq (k_1)_{G.B.}$, etc., would not offset Fogel's order of magnitude estimate of $(P_M/\overline{P})$ in equation (11).

R. E. Gallman, "Commodity Output 1839-1899" in Trends in the American Economy in the Nineteenth Century, Studies in Income and Wealth, Volume 24, NBER, Princeton: Princeton University Press, 1960, p. 26, Table 4, and Phyllis Deane and W. A. Cole, British Economic Growth 1688-1959, Cambridge: Cambridge University Press, 2nd ed., 1967, p. 166, Table 37, show that in G. B. in 1851 the distribution of commodity output by sector was 32% agriculture, 55% manufacturing, and 13% construction; and a comparable breakdown for the U. S. in 1849 shows 60% agriculture, 31% manufacturing and 9% construction. Although differences of this order persisted in the period 1840-1870, they would not by themselves serve to alter the conclusion that $(P_M/\overline{P})_{U.S.} > (P_M/\overline{P})_{G.B.}$ by the argument Fogel has applied in equation (11).

[11]Nathan Rosenberg, "Anglo-American Wage Differences in the 1820's," Journal of Economic History, Vol. XXVII, No. 2, June, 1967.

[12]Zachariah Allen, The Science of Mechanics, Providence, R. I.: Hutchens and Cory, 1829, p. 347.

[13]The American capital price could also be less than the British price since $(P_A)_{U.S.} < (P_A)_{G.B.}$ and the output of the agricultural sector is an input to the capital goods industries. Rosenberg, p. 222, Table I gives the American dollar wage of "Best Machine Makers" as $1.50-$1.75 and "Ordinary Machine Makers" as $1.25-$1.42. The corresponding rates for England are $1.94 and $1.10. Whether or not the labor cost component of capital goods is absolutely less in America depends on the proportion of the services of each category of machine maker in total labor services in capital goods production. In the argument that follows it is not necessary to establish that the American price of capital was absolutely lower than the British price.

The necessary condition is that $(P_K/\overline{P})_{U.S.} < (P_K/\overline{P})_{G.B.}$, and Allen's data definitely suggest this condition.

[14] Edward Ames and Nathan Rosenberg, "The Enfield Arsenal in Theory and History," Economic Journal, Vol. LXXVIII, No. 312, December, 1968.

[15] Ames and Rosenberg, p. 832.

[16] Although Fogel did not specify the form of the production function in the agricultural sector, this could be done, assuming homogeneous labor and that capital is used in agriculture, as follows:

$$\text{(i)} \qquad Q_A = BL^{b_1}K^{b_2}T^{b_3}, \quad (b_1+b_2+b_3) = 1$$

where,

B = index of technical efficiency in agriculture.

Corresponding to (i) the factor-price-frontier would be:

$$\text{(ii)} \qquad \frac{r}{\overline{P}} = i = g\, B^{\frac{1}{b_2}} \left(\frac{P_A}{\overline{P}}\right)^{\frac{1}{b_2}} \hat{w}^{-\frac{b_1}{b_2}} \hat{1}^{-\frac{b_3}{b_2}}$$

where,

$$g = \left(b_1^{b_1} b_2^{b_2} b_3^{b_3}\right)^{\frac{1}{b_2}}$$

$\hat{1}$ = real rental of land

$\hat{w}$ = $w/\overline{P}$, the real wage of equation (6) above.

In equation (ii) both $(P_A/\overline{P})$ and $\hat{1}$ may be used as shift parameters. Accepting the position that $(P_A/\overline{P})_{U.S.} < (P_A/\overline{P})_{G.B.}$ (tending to shift the American frontier inside to British one), the relative position of the frontiers is also governed by whether $\hat{1}_{U.S.} \gtrless \hat{1}_{G.B.}$. If contrary to the argument presented above, the real rental of land is less in the U. S. than in G. B., this would lead to indeterminateness as to the relative position of the two frontiers based on the assumption of homogeneous labor in both sectors and homogeneous capital being used in agriculture.

[17]Ian Drummond, "Labor Scarcity and the Problem of American Industrial Efficiency in the 1850's: A Comment," Journal of Economic History, Vol. XXVII, No. 3, Sept. 1967.

[18]Temin, p. 286.

[19]For a clear statement of the confusion between technological change as opposed to factor substitution within a given technology see, Nathan Rosenberg, "The Direction of Technological Change; Inducement Mechanisms and Focusing Devices," Economic Development and Cultural Change, Vol. 18, No. 1, Part 1 October 1969, pp. 1-3. As Rosenberg points out changes in relative factor prices do not explain the factor-saving bias of technological improvements within the competitive model, "The current position, as expressed by Fellner, Salter, Samuelson, and others, is that under competitive conditions on individual firm is simply not interested in the particular factor-saving bias of technical improvements."

[20]For excellent discussions of this issue in greater detail cf. Robert L. Basmann, "The Role of the Economic Historian in the Predictive Testing of 'Economic Laws,'" Explorations in Entrepreneurial History, 2d ser. Vol. II, (Spring/Summer, 1965) pp. 159-86, and J.R.T. Hughes, "Fact and Theory in Economic History," Explorations in Entrepreneurial History, 2d ser., Vol. III (Winter, 1966), pp. 75-100.

Chapter II

[1]Dorfman, Samuelson, and Solow, Linear Programming and Economic Analysis, 1958.

[2]W.E.G. Salter, Productivity and Technical Change, Monograph No. 6. Cambridge University Press, 1966.

[3]The calculations in Table 1 are based on the formula

$$-da = a\ b\ (\sigma - 1)\ \frac{dr}{r},$$

where

a = capital's share of net output
b = labor's share of net output
σ = elasticity of substitution
r = w/k = relative factor price
k = rental of capital
w = wage rate

Appendix I shows the derivation of this expression.

[4]Victor Clark, History of Manufactures in the United States. McGraw-Hill, Vol. 1, 1929, p. 392.

Nathan Rosenberg, "Anglo-American Wage Differences." Journal of Economic History, Vol. XXVII, No. 2, June, 1967, p. 226.

H. J. Habakkuk, American and British Technology in the 19th Century. Cambridge University Press, 1962, p. 76.

[5]The percentage increase in labor productivity due to substitution is given by the following equation.

$$\frac{dL}{dt} \cdot \frac{1}{L} \sigma \frac{d(w/k)}{dt} \cdot \left(\frac{k}{w}\right) \cdot \pi$$

where

$\pi = \frac{k \cdot K}{k \cdot K + wL}$ = capital's initial share of net output
L = unit labor requirements

Appendix II shows the derivation of this expression.

[6]H. J. Habakkuk, American and British Technology in the 19th Century. Cambridge University Press, 1962, pp. 49-50. "Labor-scarcity also gave Americans an incentive, not only to explore the labor-saving possibilities of autonomous inventions, but to invent new methods specifically to save labor. And if, as we shall argue later, the technical possibilities were richest at the capital-intensive end of the spectrum, the American was likely also to be better placed to make advances wherever, for any operation, he employed more capital intensive methods than the English . . ." pp. 52-53. "Where labor is scarce, any increase in productivity and real wages in one sector threatens to attract labor from other industries which either have to contract operations or install new equipment which will raise their productivity sufficiently to enable them to retain their labor supply."

[7]R. R. Nelson, "International Productivity Differences," AER, Vol. LVIII, 1969, pp. 1219-49, using a similar model found that at most one-half of the difference in labor productivity between countries could be explained by differences in the capital-labor ratio.

[8]Simon Kuznets, Capital in the American Economy, NBER, Princeton University Press, 1961, p. 209, Table 30, col. (7).

Chapter III

[1]Robert W. Fogel, Railroads and American Economic Growth. The Johns Hopkins Press, Baltimore, Md., 1964, pp. 19-21.

[2]H. W. Singer, International Development: Growth and Change. McGraw-Hill, 1964, p. 171. The specific calculation referred to by Singer was actually carried out by Agostino DeVita in the appendix to Gini's article, "Apparent and Real Causes of American Prosperity," in Banca Nazionale Del Lavoro, Rome, 1948, pp. 362-364.

[3]T. Cochran, "Did the Civil War Retard Industrialization," The Mississippi Valley Historical Review, XLVIII, September, 1961.

[4]Samuel Blodgett, Economica: A Statistical Manual for the United States of America. City of Washington, 1806, p. 58.

[5]Negro Population in the United States: 1790-1915, Bureau of the Census, U. S. Dept. of Commerce, Washington, D. C., 1918, pp. 53-54.

[6]J. Potter, "Growth of Population in America: 1700-1860," in Population in History, D. V. Glass and D.E.C. Eversley, eds., Aldine, Chicago, 1965, p. 642.

[7]Ernest Rubin, "United States," in Economics of International Migration, Brinley Thomas, ec. MacMillan and Co., Ltd., 1958, Chapter 9, p. 135.

[8]William J. Bromwell, History of Immigration to the United States. Redfield, 1856, pp. 18-19. "Finally to the 4,212,624 passengers of foreign birth arriving in the U. S. since September 30, 1819, may be added 250,000 as the number of immigrants who arrived prior to that date; making the total of foreign arrivals from the close of the Revolutionary War to December 31, 1855, 4,462,624.

[9]Herbert Heaton, "The Industrial Immigrant 1783-1812," Proceedings of the American Philosophical Society, Vol. 95, No. 5, 1951.

[10]Adam Seybert, Statistical Annals: United States of America. Philadelphia, Thomas Dobson and Son, 1818, pp. 28-31. Seybert gives an annual average estimate of foreign arrivals to the U. S. of 6,000 for the period 1790-1810. He also mentions that this estimate seems a generous one.

[11]Bromwell, op. cit., p. 15.

[12]U. S. Census of Population, 8th Census, 1860, p. xxv.

[13]Bromwell, op. cit., p. 15.

[14]Historical Statistics of the U. S. from Colonial Times to 1957, Tables C88-144 and C133-138.

[15]Walter F. Willcox, Studies in American Demography. Cornell University Press, 1940, p. 390.

[16]Bromwell, op. cit. For the period 1820-1855 the number of immigrants in the 0-5, 5-10, and 10-15 age groups are approximately equal.

[17]U. S. Census of Population, 8th Census, Mortality and Miscellaneous Statistics, 1860, p. 273, Table XXXI.

[18]A. Conrad and J. Meyer, "The Economics of Slavery in the Antebellum South," Journal of Political Economy, LXVI, No. 2, April, 1958.

[19]Conrad and Meyer, op. cit.

[20]DeBow's Review, cited in U. B. Phillips, American Negro Slavery, D. Appleton and Co., N. Y., 1918, p. 376.

[21]Kenneth Stampp, The Peculiar Institution. Alfred Knopff, New York, 1956, p. 406.

[22]Conrad and Meyer, op. cit. A maintenance cost of $10 per annum is used by the authors for slaves up to age six. Albert Fishlow, "The American Common School Revival: Fact or Fancy?" in Industrialization in Two Systems, Henry Rosovsky, ed., John Wiley and Sons, 1966, p. 55. Fishlow estimates average outlay per pupil as $4.43 in 1850 and $6.34 in 1860.

[23]W. N. Parker and F. Whartenby, "The Growth of Output before 1840," in Trends in the American Economy in the Nineteenth Century, Studies in Income and Wealth, Vol. 24, NBER, Princeton University Press, 1960. R. F. Martin, National Income in the United States, 1799-1938, National Industrial Conference Board, NICB Studies 241, 1939.

[24]Simon Kuznets, "Long Term Changes in the National Income of the United States of America from 1805 to 1950," in Income and Wealth of the United States, International Association for Research in Income and Wealth, Income and Wealth Series II, The Johns Hopkins Press, 1952.

[25]Paul David, "The Growth of Real Product in the United States before 1840: New Evidence, Controlled Conjectures," Journal of Economic History, Vol. XXVII, No. 2, June 1967, Table 8, p. 104. David reports his conjectural estimates in 1840 prices as follows:

Conjectural Estimate, Real GDP per Capita
Indexes in 1840 Prices

	Variant II	Variant I
1800	64.4	60.2
1810	61.9	57.8
1820	67.6	63.6
1830	84.0	81.8
1840	100.0	100.0

[26]David, *op. cit.*, p. 184, Table 8. The choice of David's index series gives the minimum difference between the assumed maintenance cost of $40 and per capita income in the early years of the 1790-1840 period. Martin's data implies per capita income approximately three times greater than the maintenance cost estimate at the beginning of the period.

[27]From 1790 to 1840 the Warren-Pearson Wholesale Price Index had increased from 90 to 95 for all commodities. The Bezanson index for the same period shows almost no change. We may speak of the above figures as roughly comparable real magnitudes. The annual maintenance figures are always assumed to be in current dollars.

[28]Taking the maintenance cost estimate of $40 per year as an estimate for 1840 in current prices, let us look at the social saving calculation. Valuing the

immigration flow in 1850 we will be including the social savings (maintenance) accruing in the years before 1850. Putting these estimates in the current collars of 1850 involves putting the maintenance cost estimate in 1840 of $40 in the price of 1850, or: Social Savings in 1850 = 1840 maint. cost X 1850 cost-of-living index/ 1840 cost-of-living index. In 1851, the social savings due to immigration in that year would be the 1840 maintenance cost times the ratio: 1851 cost-of-living index/ 1840 cost-of-living index, for the productive age span preceding 1851.

[29]Historical Statistics, _op. cit._, Tables E 1-12 and E 157-160, pp. 115 and 127.

[30]Specifically, regressing the cost-of-living index (Y) on the wholesale price index for the period 1820-1860 yielded the following results:

$$Y = 16.21 + 458X \; ; \quad \text{deg. of freedom} = 39$$
$$(t = 8.62) \qquad \text{adj. } r = .805$$
$$(\sigma_{yx} = .053)$$

[31]The maintenance cost figures are conceptually net of any social product contributed by children younger than the productive age. From ages 10-15 the child may not have produced a social product equal to full, social maintenance cost; but the assumption employed above implies that the difference between what a child produced and consumed from ages 10-15 was small enough to be a negligible contribution to his social maintenance cost.

[32]Corrado Gini, "Apparent and Real Causes of American Prosperity," _Banca Nazionale del Lavoro Quarterly Review_, No. 6, July 1948, p. 360.

[33]The method of estimating the capital stock is to extrapolate the trend rate of growth in the ratio of capital to value added in manufacturing ratio. This yields a ratio of 1.25 in 1860; see Col. 2, Table 5, Chapter 2; in 1879 prices or a ratio of 1.10 in 1860 prices.

$$\left(\frac{C}{\text{V.A. mfg.}}\right)_{1860} = \frac{\dfrac{C_{1879}}{CPI_{1879}}}{\dfrac{\text{V.A.}_{1879}}{VAI_{1879}}} \times \frac{\dfrac{CPI_{1879}}{CPI_{1860}}}{\dfrac{VAI_{1879}}{VAI_{1860}}}$$

The VAI, index of value added in mfg., is taken from Gallman, "Growth of Output . . .," op. cit., p. 56, Table A-5; the index of capital prices, CPI, is obtained from Gallman, "Gross National Product in the United States, 1834-1909," op. cit., p. 34, Table A-3. In 1859 Gallman gives 815 million as the value added in manufacturing, thus the capital stock is 815M x 1.1 or 895 million. In 1850 Raymond Goldsmith's estimate in Historical Statistics, Series F 197-221, shows total real tangible wealth, in 1860 prices, less government and institutional wealth, stock of specie and consumer durables of 3.9 billion dollars. The price index of goods capital formation given in Gallman, "Gross National Product in the United States, 1834-1909," op. cit., p. 34, Table A-3 was used to convert Goldsmith's current dollar estimates into the prices of 1860. According to Goldsmith's benchmark estimates in 1850 and 1880, adjusted total real tangible wealth grew at a rate of 6-1/4% per annum between those two dates. If the rate of growth of A.T.R.T.W. between 1850 and 1860 was the same as for the period 1850-1880, then this proxy for total capital stock would be equal to 7.1 billion in 1860.

[34]L. Dublin and A. Lotka, The Money Value of a Man, The Ronald Press, 1930, p. 7.

[35]Adam Smith, The Wealth of Nations, Edwin Cannan, ed. The Modern Library, 1937, Book II, Chapter 1, pp. 265-266.

[36]Alfred Marshall, Principles of Economics, 1st edition, 1890, p. 593. "In estimating the cost of the production of efficient labor, we must take as our unit the family. At all events, we cannot treat the cost of production of efficient men as an isolated problem. It must be taken as a part of the broader problem of the cost of production of efficient men together with women, who are fitted to make their homes happy, to bring up their children vigorous in body and mind,

truthful, cleanly, gentle and brave." In a following footnote Marshall has given us the rationale behind the imputation of social savings accruing from men, as well as women, in the previous section. Certain conceptual schemes would "put the value of the male immigrant too high, and that of the female immigrant too low unless allowance is made for the service which women render as mothers, as wives and/or sisters, and the male immigrants are charged with having consumed these services, while the female immigrants are credited with having supplied them."

[37]Gary Becker, Human Capital, NBER, No. 80, General Series, 1964.

[38]Corrado Gini, op. cit., p. 357. Gini has estimated also the social saving due to immigration in 1930 prices as equivalent fo 2.2 trillion dollars or six times the value of all wealth in the U.S.A. in 1930. That this calculation is seriously in error is apparent when it is considered that Gini uses 18 years as the beginning productive age. He further assumes total maintenance costs of $10,000 in 1930 prices. This latter figure he obtains from Dublin and Lotka, op. cit., pp. 57-58. This is equivalent to a maintenance cost of approximately $3,300 in the period 1820-1860 in prices of 1840, or almost ten times the maintenance cost estimate in the previous section. It also implies a maintenance cost greater than per capita income in the 1800-1840 period on my interpretation of Paul David's estimates of the trend rate of growth of gross domestic product. The Gini estimate also assumes continuous reinvestment of all the social savings at 4% per annum. To argue as Gini does that since total capital in 1930 was only one-sixth of the estimated capitalized value of social savings from 1820-1930, that America could not even invest or set aside these "gifts" at 3-1/2%-4% per annum is to gloss over the entire question of how the economy could mobilize these resources for investment. Further the economy was under no obligation to invest but could have taken out part or all of these savings for present consumption. However, if per capita disposable incomes were higher then there were implications for increased savings and investment, but not by the full amount of the social saving.

[39]While it is true that when the man dies the capital instrument is destroyed, some portion of the human capital may be transformed to physical capital permitting continuous reinvestment out of profits on the physical capital. Some portion of the human capital will go for domestic consumption and another portion for imports, thus giving rise to leakages in the continuous reinvestment process.

[40]Historical Statistics . . ., *op. cit.*, p. 61, Table C 115-132.

[41]Since Table C 115-132 shows immigration including returning citizens the gross occupational figures have been reduced to estimates of net immigration by applying the ratio that net immigration was of gross in each year to each occupational category. This ratio is shown in Column 1 of Table 3.8.

[42]While the immigrant participation rates are above those of the native population in 1840 and 1860, they were in line with the native rates in most years between 1800-1860.

[43]Historical Statistics . . ., *op. cit.*, p. 51.

[44]Lebergott, *op. cit.*, p. 467. "To measure real wage trends in the first half of the nineteenth century is probably impossible."

[45]Nathan Rosenberg, "Anglo American Wage Differences in the 1820's," The Journal of Economic History, No. 2, Vol. XXVII, June 1967, i.e., the ratio of skilled labors' wages to unskilled.

[46]C. D. Weight, *op. cit.*, p. 62. In 1795 teachers in Massachusetts were reported to earn $399.60 per year. In 1805 the low estimate was $492 per year, and the high $600. In 1816 the Superintendent of the Springfield Armory earned $700 per year. In these periods common labor earned $141, $205, and $237 per year, respectively. While these fragmentary data alone cannot sustain the conjecture, they are suggestive of the possible range of the differential. Of course these differentials are also crucially dependent upon the composition of specific occupations in each category.

[47]Bromwell, *History of Immigration to the U. S.*, *op. cit.*

Chapter IV

[1]"It is by reason of the existence of durable equipment that the economic future is linked to the present." J. M. Keynes, The General Theory of Employment, Interest and Money, Harcourt, Brace and World, 1st Harbinger edition, 1964, p. 146.

[2]The time variable may also pick up effects associated with supervisory costs and general costs of organizing the production process. It is somewhat anachronistic to think of large machines used for land clearing operation in labor abundant Latin America until one realizes the cost of recruiting and organizing the labor force along with the opportunity value of time, may make this capital intensive choice more profitable than juxtaposition of the unadorned wage rate with capital prices would indicate. At the level of conjecture, it is entirely possible that the capital intensity of production may be determined by the volume of output and the length of the production or market period rather than by elementary relative prices of the two traditional factors.

[3]Paul A. David, "The Mechanization of Reaping in the Ante-Bellum Midwest," in Industrialization in Two Systems: Essays in Honor of Alexander Gerschenkron, Henry Rosovsky, ed., John Wiley and Sons, Inc., 1966, pp. 29-30.

[4]S. B. Akers, Jr. and J. Friedman, "A Non-Numerical Approach to Production Scheduling Problems," Journal of the Operations Research Society of America, Vol. 3, No. 4, November, 1955.

[5]The illustration used here is reproduced with slight variation from Churchman, Ackoff and Arnoff, Introduction to Operations Research, John Wiley and Sons Inc., 3rd edition, 1958.

[6]H. J. Habakkuk, American and British Technology, op. cit., p. 68. Habakkuk suggested this model of the early American economy. However, I disagree with his conclusion that the profits of American manufacturers were already threatened or under pressure by the end of the 18th century; and that

the inventions of Oliver Evans and Jacob Perkins are "evidence" of this fact.

[7]The profit maximizing output depends on the position and relative difference between the MO and AO curve.

[8]George R. Taylor, The Transportation Revolution, 1815-1860, Harper and Row, 1951, Appendix A, Table 3.

[9]American State Papers, Miscellaneous, II, 286-87, Report to U. S. Senate, February 6, 1816.

[10]Eugene S. Ferguson, unpublished mimeograph on American technology.

[11]Great Britain House of Commons, Sessional Papers, 1854, Vol. 18, pp. 139, 146, 257.

[12]U. S. National Archives, Record Group 219, Report of the Second Auditor to the Second Comptroller of the Treasury. These records contain vouchers and abstracts of all labor and material purchases made by Springfield Armory.

BIBLIOGRAPHY

Government Publications

American State Papers, _Military Affairs_, 17th Congress, No. 246, "Armory at Springfield, 1795-1817."

American State Papers, _Miscellaneous_, II Report to U. S. Senate, February 6, 1816.

A Collection of Annual Reports and Other Important Papers, Relating to the Ordnance Department, prepared under the direction of Brigadier General Stephen V. Benet, 3 vols., Washington, D. C., 1878.

DeBow, J.D.B., _Compendium of the Seventh Census_, Washington, D. C., 1854.

Fitch, Charles, "Report on the Manufacture of Interchangeable Mechanism," _Tenth Census of the United States_, 1880, Vol. II, "Manufactures."

Great Britain House of Commons, _Sessional Papers_, Vol. 18, 1854.

Historical Statistics of the U. S. from Colonial Times to 1957. U. S. Department of Commerce, Bureau of the Census, 1961.

Mordecai, Alfred, _Military Commission to Europe in 1855 and 1856_, U. S. Senate Executive Documents, 36th Congress, 1st Session, 1861.

Negro Population in the United States: 1790-1915, Bureau of the Census, U. S. Department of Commerce, Washington, D. C., 1918.

Report from the Select Committee on Artizans and Machinery, Parliamentary Papers, Vol. V, 1824.

Report from the Select Committee on Small Arms, Parliamentary Papers, Vol. XVIII, 1854.

U. S. Census of Population, 8th Census, 1860.

U. S. House Documents, The Mclane Report, Serial CCXXII, 22nd Congress 1831-32, No. 308, Vol. I.

U. S. National Archives, Record Group 219, Report of the Second Auditor to the Second Comptroller of the Treasury.

Wright, Carroll D., Comparative Wages, Prices and Cost of Living, Report of the Massachusetts Bureau of the Statistics of Labor, 1885.

Books

Allen, Zachariah, The Science of Mechanics, Hutchens and Cory, Providence, R. I., 1829.

Becker, Gary, Human Capital, NBER, No. 80, General Series, 1964.

Blodgett, Samuel, Economica: A Statistical Manual for the United States of America, City of Washington, 1806.

Bromwell, William J., History of Immigration to the United States, Redfield, 1956.

Brown, Murray, On the Theory and Measurement of Technological Change, Cambridge University Press, 1966.

Churchman, C. W., Ackoff, R. L. and Arnoff, E. L., Introduction to Operations Research, 3rd ed., John Wiley and Sons, Inc., New York, 1958.

Clark, Victor, History of Manufactures in the United States, Vol. I., McGraw-Hill, 1929.

Deane, Phyllis and Cole, W. A., British Economic Growth 1688-1959, 2nd ed., Cambridge University Press, 1967.

Deyrup, Felicia, Arms Makers in the Connecticut Valley, Smith College Studies in History, Vol. XXXIII, Northhampton, Mass., 1948.

Dorfman, Samuelson, and Solow, Linear Programming and Economic Analysis, McGraw Hill, New York, 1958.

Dublin, L. and Lotka, A., Length of Life, The Ronald Press, New York, 1936.

__________, The Money Value of a Man, The Ronald Press, New York, 1930.

Ferguson, Eugene S., Early Engineering Reminiscences (1815-40) of George Eschol Sellers, Smithsonian Institution, Bulletin 238, Washington, D. C., 1965.

Fogel, Robert W., Railroads and American Economic Growth, The Johns Hopkins Press, Baltimore, 1964.

Fuller, C. E., The Whitney Firearms, Standard, Huntington, W. Va., 1946.

Habakkuk, H. J., American and British Technology in the Nineteenth Century, Cambridge University Press, 1962.

Homer, Sidney, A History of Interest Rates, Rutgers University Press, Brunswick, N. J., 1963.

Hunter, L., Steamboats on the Western Rivers, Harvard University Press, 1949.

Jewkes, Sawers and Stillerman, The Sources of Invention, St. Martin's Press, New York, 1960.

Kendrick, John W., Productivity Trends in the United States, NBER, Princeton University Press, 1961.

Keynes, J. M., The General Theory of Employment, Interest, and Money, 1st Marbinger ed., Harcourt, Brace and World, 1964.

Kuznets, Simon, Capital in the American Economy, NBER, Princeton University Press, 1961.

Lebergott, Stanley, Manpower in Economic Growth, McGraw-Hill, 1964.

MacGill, C. E., History of Transportation in the United States before 1860, Carnegie Institution of Washington, 1948.

Marshall, Alfred, Principles of Economics, 1st ed.,

Martin, R. F., National Income in the United States, 1799-1938, National Industrial Conference Board, NICB Studies 241, 1939.

Phillips, U. B., American Negro Slavery, D. Appleton and Co., New York, 1918.

Radford, G. S., Control of Quality in Manufacturing, The Ronald Press, New York, 1922.

Ringwalt, J. L., Development of Transportation Systems in the United States, Philadelphia, J. L. Ringwalt, 1888.

Roe, Joseph W., Connecticut Inventors, Yale University Press, 1934.

__________, English and American Tool Builders, Yale University Press, 1926.

Salter, W.E.G., Productivity and Technical Change, Monograph No. 6., Cambridge University Press, 1966.

Seybert, Adam, Statistical Annals: United States of America, Thomas Dobson and Son, Philadelphia, 1818.

Shewhart, Walter, Economic Control of Quality of Manufactured Product, D. van Nostrand, New York, 1931.

Singer, H. W., International Development: Growth and Change, McGraw-Hill, 1964.

Smith, Adam, The Wealth of Nations, Edwin Cannan, ed., The Modern Library, 1937.

Stampp, Kenneth, The Peculiar Institution, Alfred Knopff, New York, 1956.

Stearns, Charles, The National Armories, Springfield, Mass., 1852.

Swank, J. M., Progressive Pennsylvania, J. B. Lippincott, Philadelphia, 1908.

Taylor, George R., The Transportation Revolution, 1815-1860, Harper and Row, 1951.

Usher, Abbott Payson, The History of Mechanical Invention, Beacon Press, Boston, 1959.

Waters, Asa H., Biographical Sketch of Thomas Blanchard and His Inventions, Worcester, Mass., 1878.

Willcox, Walter F., Studies in American Demography, Cornell University Press, 1940.

Woodbury, Robert, History of the Milling Machine, Technology Press, M.I.T., Cambridge, Mass., 1960.

Young, Edward, Labor in Europe and America, Philadelphia, 1875.

Articles

Akers, S. B., Jr. and Friedman, J., "A Non-Numerical Approach to Production Scheduling Problems," Journal of the Operations Research Society of America, Vol. 3, No. 4, November 1955.

Ames, Edward and Rosenberg, Nathan, "The Enfield Arsenal in Theory and History," Economic Journal, Vol. LXXVIII, No. 312, December 1968.

Arrow, K. et al, "Investment and Technical Progress," in Mathematical Methods in the Social Sciences, Stanford University Press, 1959.

Basmann, Robert L., "The Role of the Economic Historian in the Predictive Testing of 'Economic laws,'" Explorations in Entrepreneurial History, 2nd series, Vol. II, Spring/Summer 1965.

Budd, Edward C., "Factor Shares, 1850-1910," in _Trends in the American Economy in the Nineteenth Century_, Studies in Income and Wealth, Vol. 24, NBER, Princeton University Press, 1960.

Burn, D. L., "The Genesis of American Engineering Competition, 1850-1870," _Economic History_, January 1931.

Cochran, T., "Did the Civil War Retard Industrialization?" _The Mississippi Valley Historical Review_, XLVIII, September 1961.

Colt, Samuel, "On the Application of Machinery to the Manufacture of Rotating Chambered-Breech Fire-Arms, and the peculiarities of those arms," _Minutes of Proceedings of the Institution of Civil Engineers_, Vol. XI, Session 1851-52.

Conrad, A. and Meyer, J., "The Economics of Slavery in the Antebellum South," _Journal of Political Economy_, LXVI, No. 2, April 1958.

David, Paul, "The Growth of Real Product in the United States before 1840: New Evidence, Controlled Conjectures," _Journal of Economic History_, Vol. XXVII, No. 2, June 1967.

David, Paul, "The Mechanization of Reaping in the Ante-Bellum Midwest," in _Industrialization in Two Systems: Essays in Honor of Alexander Gerschenkron_, Henry Rosovsky, ed., John Wiley and Sons, Inc., 1966.

Drummond, Ian, "Labor Scarcity and the Problem of American Industrial Efficiency in the 1850's: A Comment," _Journal of Economic History_, Vol. XXVII, No. 3, September 1967.

Durfee, W. F., "The First Systematic Attempt at Interchangeability in Firearms," _Cassier's Magazine_, Vol. V, April 1894.

Ferguson, C. E. and Pfouts, P. W., "Aggregate Production Functions and Relative Factor Shares," _International Economic Review_, Vol. 3, September 1962.

Ferguson, Eugene S., unpublished mimeograph on American technology.

Fishlow, Albert, "The American Common School Revival: Fact or Fancy?" in Industrialization in Two Systems, Essays in Honor of Alexander Gerschenkron, Henry Rosovsky, ed., John Wiley and Sons, Inc., 1966.

Fogel, Robert W., "The Specification Problem in Economic History," Journal of Economic History, Vol. XXVII, No. 3, September 1967.

__________, "The Specification Problem in Economic History, A Correction," Journal of Economic History, Vol. XXVII, No. 7, March 1968.

Gallman, Robert E., "Commodity Output in the United States, 1839-1899," in Trends in the American Economy in the Nineteenth Century, Studies in Income and Wealth, Vol. 24, NBER, Princeton University Press, 1960.

__________, "Gross National Product in the United States, 1834-1909," in Output, Employment and Productivity in the U. S. after 1800, Studies in Income and Wealth, Vol. 30, NBER, Princeton University Press, 1966.

Gini, Corrado, "Apparent and Real Causes of American Prosperity," Banca Nazionale del Lavoro Quarterly Review, No. 6 (Rome), July 1948.

Heaton, Herbert, "The Industrial Immigrant, 1783-1812," Proceedings of the American Philosophical Society, Vol. 95, No. 5, 1951.

Hubbard, Guy, "Development of Machine Tools in New England," American Machinist, Vol. LX, February 21, 1924.

Hughes, J.R.T., "Fact and Theory in Economic History," Explorations in Entrepreneurial History, 2nd series, Vol. III, Winter 1966.

Kuznets, Simon, "Long Term Changes in the National Income of the United States of America from 1805 to 1940," in Income and Wealth of the United

States, International Association for Research in Income and Wealth, Income and Wealth Series II, The Johns Hopkins Press, 1952.

Kuznets, S. and Rubin, E., "Immigration and the Foreign Born," NBER, Occ. Paper No. 46, 1954.

Lebergott, Stanley, "Labor Force and Employment, 1800-1960," in Output, Employment and Productivity in the U. S. after 1800, Studies in Income and Wealth, Vol. 30, NBER, Princeton University Press, 1966.

Leontief, W. W., "Domestic Production and Foreign Trade: The American Capital Position Re-Examined," Proceedings of the American Philosophical Society, XCVII, September 1953.

__________, "Factor Proportions and the Structure of American Trade: Further Theoretical and Empirical Analyses," Review of Economics and Statistics, XXXVII, November 1956.

Lorant, John W., "Technological Change in Manufacturing," Journal of Economic History, XXVII, Number 2, June 1967.

Meier, Hugo A., "American Technology and the Nineteenth-Century World," American Quarterly, Vol. 10, 1958.

Nelson, R. R., "International Productivity Differences," American Economic Review, Vol. LVIII, Number 5, December, 1968.

Orcutt, H.F.L., "Modern Machine Methods," Proceedings of the Institution of Mechanical Engineers, Parts 1-2, 1902.

Parkhurst, E. G., "Origin of the Turret, or Revolving Head," American Machinist, Vol. XXIII, May 24, 1900.

Potter, J., "Growth of Population in America: 1700-1860," in Population in History, D. V. Glass and D.E.C. Eversley, ed., Aldine, Chicago, 1965.

Roe, Joseph W., "History of the First Milling Machine," American Machinist, Vol. XXXVI, June 27, 1912.

__________, "Interchangeable Manufacture," Transactions of the Newcomen Society, Vol. XVIII, 1937.

Rosenberg, Nathan, "Anglo-American Wage Differences in the 1820's," Journal of Economic History, Vol. XXVII, No. 2, June 1967.

__________, "The Direction of Technological Change: Inducement Mechanisms and Focusing Devices," Economic Development and Cultural Change, Vol. 18, No. 1, Part 1, October 1969.

__________, "Technological Change in the Machine Tool Industry, 1840-1910," Journal of Economic History, Vol. XXIII, Number 4, December 1963.

Rothbarth E., "Causes of the Superior Efficiency of U.S.A. Industry as Compared with British Industry," Economic Journal, Vol. LVI, September 1946.

Rubin, Ernest, "United States," in Economics of International Migration, Brinley Thomas, ed., MacMillan and Co., Ltd., 1958.

Samuelson, P. A., "A New Theorem on Nonsubstitution," in Money, Growth and Methodology, Hugo Hegeland, ed., Social Science Studies, No. 20, Lund, 1961.

__________, "Parable and Realism in Capital Theory, The Surrogate Production Function," Review of Economic Studies, Vol. XXIX, No. 30, June 1962.

__________, "A Theory of Induced Innovation Along Kennedy-Weisacker Lines," Review of Economics and Statistics, Vol. XLVII, November, 1965.

Sawyer, John, "The Social Basis of the American System of Manufacturing," Journal of Economic History, Vol. 14, No. 4, December 1954.

Simon, Herbert, "Effects of Technological Change in a Linear Model," in Activity Analysis of Production and Allocation, T. C. Koopmans, ed., John Wiley, New York, 1951.

Spotts, M. F., "The Statistical Dimensioning of Machine Parts," American Society of Mechanical Engineers, Transactions, 1958.

Stigler, George, "Production and Distribution in the Short Run," Journal of Political Economy, Vol. 47, June 1939.

Temin, Peter, "Labor Scarcity and the Problem of American Industrial Efficiency in the 1850's," Journal of Economic History, Vol. XXVI, No. 3, September 1966.

Dissertations in American Economic History

An Arno Press Collection

Adams, Donald R., Jr. **Wage Rates in Philadelphia, 1790-1830.** (Doctoral Dissertation, University of Pennsylvania, 1967). 1975

Aldrich, Terry Mark. **Rates of Return on Investment in Technical Education in the Ante-Bellum American Economy.** (Doctoral Dissertation, The University of Texas at Austin, 1969). 1975

Anderson, Terry Lee. **The Economic Growth of Seventeenth Century New England:** A Measurement of Regional Income. (Doctoral Dissertation, University of Washington, 1972). 1975

Bean, Richard Nelson. **The British Trans-Atlantic Slave Trade, 1650-1775.** (Doctoral Dissertation, University of Washington, 1971). 1975

Brock, Leslie V. **The Currency of the American Colonies, 1700-1764:** A Study in Colonial Finance and Imperial Relations. (Doctoral Dissertation, University of Michigan, 1941). 1975

Ellsworth, Lucius F. **Craft to National Industry in the Nineteenth Century:** A Case Study of the Transformation of the New York State Tanning Industry. (Doctoral Dissertation, University of Delaware, 1971). 1975

Fleisig, Heywood W. **Long Term Capital Flows and the Great Depression:** The Role of the United States, 1927-1933. (Doctoral Dissertation, Yale University, 1969). 1975

Foust, James D. **The Yeoman Farmer and Westward Expansion of U. S. Cotton Production.** (Doctoral Dissertation, University of North Carolina at Chapel Hill, 1968). 1975

Golden, James Reed. **Investment Behavior By United States Railroads, 1870-1914.** (Doctoral Thesis, Harvard University, 1971). 1975

Hill, Peter Jensen. **The Economic Impact of Immigration into the United States.** (Doctoral Dissertation, The University of Chicago, 1970). 1975

Klingaman, David C. **Colonial Virginia's Coastwise and Grain Trade.** (Doctoral Dissertation, University of Virginia, 1967). 1975

Lang, Edith Mae. **The Effects of Net Interregional Migration on Agricultural Income Growth:** The United States, 1850-1860. (Doctoral Thesis, The University of Rochester, 1971). 1975

Lindley, Lester G. **The Constitution Faces Technology:** The Relationship of the National Government to the Telegraph, 1866-1884. (Doctoral Thesis, Rice University, 1971). 1975

Lorant, John H[erman]. **The Role of Capital-Improving Innovations in American Manufacturing During the 1920's.** (Doctoral Thesis, Columbia University, 1966). 1975

Mishkin, David Joel. **The American Colonial Wine Industry:** An Economic Interpretation, Volumes I and II. (Doctoral Thesis, University of Illinois, 1966). 1975

Oates, Mary J. **The Role of the Cotton Textile Industry in the Economic Development of the American Southeast:** 1900-1940. (Doctoral Dissertation, Yale University, 1969). 1975

Passell, Peter. **Essays in the Economics of Nineteenth Century American Land Policy.** (Doctoral Dissertation, Yale University, 1970). 1975

Pope, Clayne L. **The Impact of the Ante-Bellum Tariff on Income Distribution.** (Doctoral Dissertation, The University of Chicago, 1972). 1975

Poulson, Barry Warren. **Value Added in Manufacturing, Mining, and Agriculture in the American Economy From 1809 To 1839.** (Doctoral Dissertation, The Ohio State University, 1965). 1975

Rockoff, Hugh. **The Free Banking Era: A Re-Examination.** (Doctoral Dissertation, The University of Chicago, 1972). 1975

Schumacher, Max George. **The Northern Farmer and His Markets During the Late Colonial Period.** (Doctoral Dissertation, University of California at Berkeley, 1948). 1975

Seagrave, Charles Edwin. **The Southern Negro Agricultural Worker:** 1850-1870. (Doctoral Dissertation, Stanford University, 1971). 1975

Solmon, Lewis C. **Capital Formation by Expenditures on Formal Education, 1880 and 1890.** (Doctoral Dissertation, The University of Chicago, 1968). 1975

Swan, Dale Evans. **The Structure and Profitability of the Antebellum Rice Industry:** 1859. (Doctoral Dissertation, University of North Carolina at Chapel Hill, 1972). 1975

Sylla, Richard Eugene. **The American Capital Market, 1846-1914:** A Study of the Effects of Public Policy on Economic Development. (Doctoral Thesis, Harvard University, 1968) 1975

Uselding, Paul John. **Studies in the Technological Development of the American Economy During the First Half of the Nineteenth Century.** (Doctoral Dissertation, Northwestern University, 1970) 1975

Walsh, William D[avid]. **The Diffusion of Technological Change in the Pennsylvania Pig Iron Industry, 1850-1870.** (Doctoral Dissertation, Yale University, 1967). 1975

Weiss, Thomas Joseph. **The Service Sector in the United States, 1839 Through 1899.** (Doctoral Thesis, University of North Carolina at Chapel Hill, 1967). 1975

Zevin, Robert Brooke. **The Growth of Manufacturing in Early Nineteenth Century New England.** 1975